ObjectARX™ Primer

ObjectARX™ Primer

Bill Kramer

Press

Thomson Learning™

Africa • Australia • Canada • Denmark • Japan • Mexico • New Zealand
Phillipines • Puerto Rico • Singapore • Spain • United Kingdom • United States

NOTICE TO THE READER

Trademarks

Autodesk Press Staff
Executive Director: Alar Elken
Executive Editor: Sandy Clark
Development/Editorial Assistant: Allyson Powell
Executive Marketing Manager: Maura Theriault
Executive Production Manager: Mary Ellen Black
Production Coordinator: Jennifer Gaines
Art and Design Coordinator: Mary Beth Vought
Marketing Coordinator: Paula Collins
Technology Project Manager: Tom Smith

Cover illustration by Brucie Rosch

For more information, contact
Autodesk Press
3 Columbia Circle, Box 15-015
Albany, New York USA 12212-15015;
or find us on the World Wide Web at
http://www.autodeskpress.com

Library of Congress Cataloging-in-Publication Data

Kramer, Bill, 1958–
    ObjectARX primer / Bill Kramer.
        p.   cm.
    ISBN 0–7668–1127–1 (pbk.)
    1. Object-oriented programming (Computer science) 2. ObjectARX.
    3. AutoCAD (Computer file) I. Title.
    QA76.64.K73 1999
604.2'0285'5117--dc21   99–32454
                                                                        CIP

CONTENTS

CHAPTER 9 BUILDING AN OBJECTARX APPLICATION

FOREWORD

A WORD FROM THOMSON LEARNING™

Autodesk Press was formed in 1995 as a global strategic alliance between Thomson Learning and Autodesk, Inc. We are pleased to bring you the premier publishing list of Autodesk student software and learning and training materials to support the Autodesk family of products. AutoCAD® is such a powerful product that everyone who uses it would benefit from a mentor to help them unlock its full potential. This is the premise upon which the Programmer Series was conceived. The titles in this series cover the most advanced topics that will help you maximize AutoCAD. Our Programmer Series titles also bring you the best and the brightest authors in the AutoCAD community. Maybe you've read their columns in a CAD journal, maybe you've heard them speak at an Autodesk event, or maybe you're new to these authors —whatever the case may be, we know you'll enjoy and apply what you'll learn from them. We thank you for selecting this title and wish you well on your programming journey.

Sandy Clark
Executive Editor
Autodesk Press

A WORD FROM AUTODESK, INC.

From the birth of AutoCAD onward, there has been a large library of source material on how to use the software; however, relatively little material on customization of AutoCAD has been made available. There have been a few texts written about AutoLISP®, and many general AutoCAD books include a chapter or two on customization. On the whole, however, the number of texts on programming AutoCAD has been inadequate given the amount of open technology, the number of application programming interfaces (APIs), and the sheer volume of opportunity to program the AutoCAD design system.

Four years ago, when I started working with developers as the product manager for AutoCAD APIs, I immediately found an enormous demand for supporting texts about ObjectARX™, AutoLISP, Visual Basic for Applications®, and AutoCAD® OEM. The demand for a "programming series" of books stems from AutoCAD's history of being the most programmable, customizable, and extensible design system on the market. By one measure, over 70% of the AutoCAD customer base "programs" AutoCAD using Visual LISP™, VBA, or menu customization, all in an effort to increase productivity. The demand for increasing the designer's productivity extends deeply, creating a demand for new source texts to increase productivity of the programming and customization process itself.

The standard Autodesk documentation for AutoCAD provides the original source of technical material on the AutoCAD API technology. However, it tends toward the clinical, which is a natural result of describing the software while it is being created in the software development lab. Documentation in fully applied depth and breadth is only completed through the collective experiences of hundreds of thousands of developers, customers, and users as they interpret and apply the system in ways specific to their needs.

As a result, demand is high for other interpretations of how to use AutoCAD APIs. This kind of instruction develops the technique required to innovate. It develops programmer instinct by instructing when to use one interface over another and provides direction for interpretive nuances that can only be developed through experience. AutoCAD customers and developers look for shortcuts to learning and for alternative reference material. Customers and developers alike want to accelerate their programming learning experience, thereby shortening the time needed to become expert and enabling them to focus sooner and better on their own specific customization or development projects. Completing a customization or development project sooner, faster, and better means greater productivity during the development project and more rapid deployment of the result.

For the CAD manager, increasing productivity through accelerated learning means increasing his or her CAD department's productivity. For the professional developer, this means bringing applications to market faster and remaining competitive.

THE INITIAL IDEA

An incident that occurred at Autodesk University in Los Angeles in the fall of 1997 illustrated dramatically a dynamic demand for AutoCAD API technology information. Bill Kramer was presenting an overview session on AutoCAD's ObjectARX. The Autodesk University officials had planned on having 20 to 30 registrants for this session and had assigned an appropriately sized room to the session. About three weeks before it took place, I received a telephone call indicating that the registrations for the session had reached the capacity of the room, and we would be moving the session.

We moved the session two more times due to increasing registration from customers, CAD Managers, designers, corporate design managers, and even developers attending the Autodesk University customer event. What they all had in common was an interest in seeing how ObjectARX, an AutoCAD API, was going to increase their own or their department's design productivity. When the presentation started, I walked into the back of the room to see what appeared to me to be over 250 people in a room which was now, standing room only. That may have been the moment when I decided to act, or it may have been just when the actual intensity of this demand became apparent to me; I'm not sure which.

The audience was eager to hear what Bill Kramer was going to say about the power of ObjectARX. Nearly an hour of unexpected follow-up questions and answers followed. Bill had successfully evangelized a technical subject to an audience spanning non-technical to technical individuals. This was a revelation for me, and the beginning of my interest in developing new ways to communicate to more people the technical aspects, power, and benefits of AutoCAD technology and APIs.

As a result, I asked Bill if he would write a book on the very topic he just presented. I'm happy to say that Bill's book has become one of the first books written in the new AutoCAD Programmer Series with Autodesk Press.

TOPICS FOR THE PROGRAMMER SERIES

The extent of AutoCAD's open programmable design system made the need for a series of books apparent early on. My team, under the leadership of Cynde Hargrave, Senior AutoCAD Marketing Manager, began working with Autodesk Press to develop this series.

It was an exciting project, with no shortage of interested authors covering a range of topics from AutoCAD's open kernel in ObjectARX to the Windows standard for application programming in VBA. The result is a complete library of references in Autodesk's Programmer Series covering ObjectARX, Visual LISP, AutoLISP, customizing AutoCAD through ActiveX Automation® and Microsoft Visual Basic for Applications, AutoCAD database connectivity, and general customization of AutoCAD.

WHO READS PROGRAMMING BOOKS?

Every AutoCAD user will find books in this series to fit their AutoCAD customization or development interests. The collective goal we had with our team and Autodesk Press for developing this series, identifying titles, and matching them with authors was to provide a broad spectrum of coverage across a wide variety of customization content and a wide range of reader interest and experience.

Collaborating as a team, Autodesk Press and Autodesk developed a programmer series covering all the important APIs and customization topics. In addition, the series provides information that spans use and experience levels from the novice just starting to customize AutoCAD to the professional programmer or developer looking for another interpretive reference to increase his or her experience in developing powerful applications for AutoCAD.

THIRST, THEME, AND VARIATION

I compare this thirst for knowledge with the interest musicians have in listening to music performed in different ways. For me, it is to hear Vivaldi's The Four Seasons time and time again. Musicians play from the same notes written on the page, with the identical crescendos and decrescendos and other instructions describing the "technical" aspects of the music.

All of the information to play the piece is there. However, the true creative design and beauty only manifests through the collection of individual musicians, each applying a unique experience and interpretation based on all that he or she has learned before from other mentors in addition to his or her own practice in playing the written notes.

By learning from other interpretations of technically identical music, musicians benefit the most from a new, and unique, interpretation and individual perception. This makes it possible for musicians to amplify their own experience with the technical content in the music. The result is another unique understanding and personalized interpretation of the music.

Similar to musical interpretation is the learning, mentoring, creative processes, and resources required in developing great software, programming AutoCAD applications, and customizing the AutoCAD design system. This process results in books such as Autodesk's Programmer Series, written by industry and AutoCAD experts who truly love working with AutoCAD and personalizing their work through development and customization experience. These authors, through this programmer series, evangelize others, enabling them to gain from their own experiences. For us, the readers, we gain the benefit from their interpretation, and obtain the value through different presentation of the technical information, by this wide spectrum of authors.

Andrew Stein
Senior Manager
Autodesk Business Research, Analysis and Planning

PREFACE

IS THIS BOOK FOR YOU?

If you are involved with AutoCAD applications, as a user or as a developer this book is for you. You do not need to be a programmer to understand what this book teaches. You do not even need to work with programmers directly to benefit from these contents. ObjectARX is a tool kit for programmers; however, it is also a new way of looking at problem solving through customization in the CAD/CAM/CAE systems arena.

You should read this book if:

- You are developing AutoCAD applications using any of the existing tools.
- You are managing the development of AutoCAD applications.
- You are a user of AutoCAD looking for more productivity.
- You are curious as to what ObjectARX is and what it can do for you.
- You are considering the development of a new AutoCAD based application.
- You are interested in learning more about applications of object-oriented programming.

WHY WE WROTE THIS BOOK

When new things are created it is sometimes hard to describe them. It takes some time for people to adjust in their thinking habits about how things are supposed to be and work. Until then, attempts at describing the new things and how or why to use them may be difficult at best. In some cases there can even be cultural issues that directly affect the way things are perceived and how things should be done. New tools that require specialized operators are seen as expensive and not worth investing in until proven by someone else to be good.

ObjectARX has been like that. It is not entirely new and the existing information on it is very technical in nature. Describing ObjectARX to someone who does not understand the concepts can be difficult. Learning it can be even more so without some background concepts well under control.

That is one of the reasons why this book was written. To introduce the power and capabilities of ObjectARX to programmers and programming managers of AutoCAD application projects. Because of the depth of ObjectARX this book was needed to help build a bridge from one way of thinking about CAD applications to another way of thinking altogether. ObjectARX is an object-oriented programming tool for interfacing with AutoCAD at a very intimate level. Unless all levels of development (user, manager and programmer) understand the capabilities then it cannot be exploited properly. Therefore this book was written with the idea of providing managers and programmers of AutoCAD applications a tool by which they can both learn about the tool kit and communicate about it while the projects are underway.

There is a great deal of curiosity about ObjectARX amongst AutoLISP and VBA programmers looking for even more powerful programming tools inside AutoCAD. This book was also written for them to provide a guiding light as to how ObjectARX can be used to augment existing applications. Even if one is only considering the use of Visual LISP and not ObjectARX with C++, the understanding of objects and how they are related to programming CAD applications is valuable information. Thus the second reason for writing this book was to introduce existing AutoCAD programmers, who do not currently know C++, to the most powerful tool for customizing AutoCAD.

A third reason behind this book is to teach the concepts of object-oriented programming to those who have not been exposed to such concepts before. Since engineers and architects tend to write AutoCAD applications, there is a strong chance they have never been directly exposed to object-oriented programming in the past. The concepts behind object-oriented programming are not difficult but they do require some time to explain. This book takes the time needed to explain the basics of objects and how they represent a new way of thinking about programming AutoCAD applications. A historical context is used to show how the pieces all fit together to deliver this most powerful tool called ObjectARX.

As just seen, this book was written for multiple perspectives. Although most readers will have some background in programming it is not required in order to gain value from this work. The original intent was to provide a manager level explanation of the technology, an overview so to speak. As such, engineers and architects involved in the management of AutoCAD applications development will benefit greatly. Programmers eager to learn the ObjectARX system, but put off by the seemingly deep technical jargon used, can also use this work as a jumping point to a greater understanding of how ObjectARX works.

If you have ever wondered if ObjectARX could be used to enhance your applications, then this book is a great starting place to learn about the technology and what it can do for you. I wish I had it when I was first learning about ObjectARX.

THE STRUCTURE OF THIS BOOK

The chapters in this book are organized to provide an increasingly deeper technical picture into the ObjectARX environment without getting lost in programming examples. Basic concepts are discussed first and then the ObjectARX system components are introduced. ObjectARX for AutoCAD Release 14 and ObjectARX for AutoCAD 2000 are very much the same at the level we are presenting this information. Specific differences are called out in the text as the more technical aspects of the information being covered are presented.

The first chapter begins by explaining why you should want to customize AutoCAD and how to justify the activity. Too many users of AutoCAD are working with the product out of the box and that is not the best way to gain a significant return on investment. CAD/CAM/CAE systems are intended to improve productivity and accuracy, but they are not the final solution in themselves. On-going improvements through customization can provide powerful tools for designers to use.

The second chapter explains object-oriented programming concepts using simple conceptual models. Object-oriented programming is not a difficult concept to understand but it is a different way of thinking for many programmers. This chapter explores what objects are about and the theory as to how they are implemented in a computer.

These concepts are brought together in chapter three where a historical context is used to show objects and CAD systems were meant for each other. The rapid evolution of computer technology and software has left many wondering about the tools they are using and where things are going next. As seen in many other technologies, a look at the history behind the components can often provide insight into what will happen next.

Chapter four is where the AutoCAD object-oriented programming environment is first introduced in some detail. After taking a brief look at the libraries provided and what they represent, a more detailed description of how an ObjectARX application fits into the AutoCAD system is presented. The remaining libraries are then revealed in more detail with explanations of how they fit into applications in general.

Entity objects are the main topic covered in chapter five. How entity objects are addressed in the system is very important in programming applications for AutoCAD. A detailed look at the processing requirements and techniques involved in manipulating entities provides a jumping point into developing entity-based applications with ObjectARX. Some C++ code is introduced to demonstrate how easily ObjectARX creates and manipulates objects.

A new concept to many AutoCAD applications programmers involves reactors. The concepts and applications of reactors are discussed in chapter six. How reactors work inside the computer at a conceptual level is important to understand when developing programs that utilize this feature. There are several different kinds of reactors available inside ObjectARX, and this chapter briefly introduces each type.

Chapter seven goes into another powerful and new feature of ObjectARX, the creation of new entity objects. This is an area where there has been a great deal of confusion. The process involved and the details that need to be addressed are covered so that applications programmers and managers alike know what it takes to make custom objects.

Object support does extend beyond ObjectARX to a limited degree and chapter eight looks at the differences of the various programming tools available. Sample application ideas are presented and discussed in general terms to show how these various tools all fit together to form a very powerful programming environment for engineering and architectural applications development.

The last chapter provides a step by step introduction to writing your first ObjectARX program using Microsoft Visual C++. A detailed look at the steps needed to create, to compile, and to build a working ObjectARX program are shown with screen captures to help guide the beginning C++ user. Because of the differences found in each, both AutoCAD Release 14 and AutoCAD 2000 versions are presented in the example.

ACKNOWLEDGMENTS

I'd like to thank Autodesk for this wonderful tool kit and even more, for the opportunity to present it in written form to all interested. The developers behind ObjectARX are to be commended for having the foresight to see the technology and for having the fortitude required to bring it to reality. Autodesk provided the brunt of the editing work and technical guidance, and for that I am grateful. Specifically, I'd like to thank Andrew and Cynde for the great job they did in making this work possible.

I'd also like to thank Nancy for serving as a sounding board. She didn't understand object-oriented programming when we started, but she does now. In addition, I'd like to thank my family for continuing to put up with my sometimes-eccentric antics in both science and computers, including the starting of this book while on vacation. Lastly I'd like to thank my Dad for always challenging me to think about things in a different way in order to solve problems creatively.

This book was a lot of fun to write and I hope it is equally interesting to read. It was a test to present ObjectARX without too many programming examples and without getting lost in the computer science jargon. I am glad I took on the project as I learned a great deal by being forced to rethink the environment from multiple perspectives. As a computer scientist who has been involved in CAD/CAM/CAE for two decades, this technology has been a welcomed tool and has been part of a logical progression of the technology. I am quite pleased to be able to share my insights into this exciting tool and how it fits into the scheme of CAD/CAM/CAE operations.

The future is very bright for CAD/CAM/CAE applications development. And ObjectARX presents a technology that can take it to an even higher level. There is a lot of work to be done creating even more objects and that will keep programmers busy well into the next century.

Keep on programmin'!

Bill Kramer, April 1999

CHAPTER 1

Customization of AutoCAD®

The single most important feature ever added to AutoCAD was the ability to be customized. No two people do the same thing the same way all the time. Through skilled customization AutoCAD is transformed into a very powerful tool for the design professional. Unfortunately many users of AutoCAD have not given the thought of customization sufficient time and energy and are forced to run the software as provided in the box. The goal of this chapter is to show where to put time and energy into customization, and then how to justify the efforts involved.

TRAINING AUTOCAD

AutoCAD is a drafting tool used in the creation of engineering and architectural drawings, much like a drafting table and a T-square. Anyone skilled in the art of using AutoCAD software can create drawings of virtually any description, just the same as a skilled draftsman can create a drawing given the instructions. Of course, an experienced draftsman can create a drawing faster than a draftsman who is new to a given discipline of drawing can. The experienced draftsman also requires less instruction as the past experiences combine to provide an enhanced communication between the designer and the drawer. This is why experienced drafters are paid more.

Just as the experienced draftsman can create a drawing with fewer instructions and in less time than a novice draftsman, an AutoCAD system that has been trained in a specific discipline can create a drawing with fewer instructions and in less time.

A trained AutoCAD system can take on many forms ranging from a modified menu structure to a system that runs AutoCAD transparently. In some cases it's hard to identify what has been modified from the original AutoCAD, and users may think they are running the original system. In fact, a user trained in generic AutoCAD can become more proficient in a given discipline with a highly trained system. But this is only if the training is done right, both for the user and the computer.

AutoCAD users must still be skilled in the usage of AutoCAD in order to achieve maximum productivity. The more experience they have in the discipline involved, the better. The possible productivity gains through automation are tremendous, especially when compared with manual methods. It's possible to create new, complete design drawings in minutes and even seconds given the right conditions. Even if the conditions do not merit such astounding achievements in productivity, there can be significant cost savings in the documentation process of a design with a trained AutoCAD system.

Computer systems are trained through customization. Customization happens at multiple levels inside AutoCAD and, for the most part, involves programming. At each level, different programming skills are required to accomplish the customization. The actual time required to create the custom module will vary depending on the complexity of the project and the skill of the programmer. For example, creating 100 text note detail drawings for insertion into drawings as blocks and then making menu entries for each of them will most likely take longer than writing a program to insert text files into the drawing.

THE TOP REASONS TO CUSTOMIZE AUTOCAD

The following are the main reasons to customize AutoCAD. They are not in any particular order as one may be more important than another to someone in a given situation.

FURTHER ENHANCE PRODUCTIVITY

The primary reason to customize AutoCAD in all drawing environments is to improve the productivity of the operation. The goal is to reduce the time and cost required to obtain drawings created from designs.

Providing tools that the operators need to accomplish the tasks at hand enhances productivity. Automated dimensioning, pre-drawn components, detailing tools, and data import programs are just of a few of the customization paths to investigate that can have an impact on productivity. The complexity of the tools is directly related to the complexity of the tasks to be solved and the skill levels of the operators.

IMPROVE QUALITY

Improving the overall quality of the work can be an important reason for customization in larger drawing shops. Some operations may make use of temporary or "transient" CAD operators to complete large projects. In these cases, productivity tools that enforce standards will yield tremendous benefits. Not only will the drawings be created a bit faster, but they will also be created to a specific standard of quality.

WORK THE WAY YOU WORK

Changing AutoCAD to work the way you need it to work is another good reason to customize the system. This may be considered a parallel concept to improving productivity, as the two tend to achieve the same goal. What motivates the user is different; this can become important when training new people to use the system, or when working with occasional users of the CAD/CAM/CAE system.

To illustrate the point, there are many instances—such as survey notes—where the data available requires additional reduction before it can be used in AutoCAD. A data reduction system is a perfect example of a customization to AutoCAD to make it work the way you work. Not only can the data be entered into AutoCAD faster, but in a fashion that the user is comfortable with in the first place.

Another example is to change the language of the system to talk the way you talk. A drill hole object which combines various elements of the basic drawing package (circles, lines and text) to allow you to think in terms of drill holes and not the individual components would significantly impact productivity and training requirements.

SOLVE GRAPHICAL PROBLEMS

Using the computer to solve mathematical problems is always a good use of the tool. For many engineering and scientific calculations the output can best be rendered graphically. AutoCAD provides the perfect vehicle for this sort of application. Not only can the results be shown in a graph; the graph can be created in a drawing suitable for immediate plotting.

In this case, AutoCAD serves primarily as an output engine. The goal is to create graphical information automatically and have plots generated from these outputs. Examples include drawing performance curves based on experimental results of a motor in operation, showing geological information based on core samples, and drafting statistical charts from various source data. In all these situations we are basically reducing the data and then plotting a drawing.

ENGINEERING COMPUTATIONS

Engineering mathematics is often specific to each unique application or environment. It's common to have tables of pre-calculated results handy for the most common designs. Having a tool that calculates these values inside the CAD system will not only save time but also improve accuracy of the calculations by removing the need to enter numbers in the calculator.

For engineering computations, AutoCAD serves both as an input and output engine depending on the nature of the application. Problems such as highway alignment offsets to curb locations, reinforcing steel layouts, electronic circuit analysis, and moment of inertia are examples of engineering computations that can involve both graphical input and output.

PROVIDE DATA INTEGRATION

Bringing isolated computer systems together is integration. There are many aspects to true CAD/CAM/CAE integration, but the essence behind it means not putting information into the computer more than once. This practice reduces potential errors by reducing human input to a bare minimum and can improve productivity.

One potential problem with integration is that the productivity improvement may not be at the AutoCAD workstation all the time. As an example, a program may exist which works with 3D models after they have been created. Now suppose the models must conform to a specific standard that, if done with AutoCAD out-of-the-box commands, would be cumbersome to the operators. Providing additional tools which make conforming to the requirements easier will greatly enhance the overall system. Even though the tools may not be something that would normally be developed, the overall system improves.

A bill of materials extraction into a spreadsheet for costing analysis or billing is a classic example of this sort of integration. Another example is storing features about some object in another data system as in a mapping application where there may be a great deal of non-graphical information associated with an object.

The important thing to remember is that only through customization can AutoCAD be integrated with other productivity tools on your computer. This means that if you want AutoCAD to help generate a spreadsheet or database or to help tie into the machine tools, you will need to customize the system.

THE POTENTIAL BENEFITS OF CUSTOMIZATION

It's hard to measure in a direct equation the benefits that can be derived by customizing the AutoCAD system. This is mostly because they vary greatly from case to case and obtaining concrete numbers from which to base an analysis can be difficult. In some situations, the time spent gathering information about how much time could be saved is many times greater than the cost of just doing the programming. For the most part, one must rely on the input of the operators to determine where to apply customization efforts to get the greatest benefits for a specific operation.

The user audience also plays the most important role in the benefits analysis when timesaving is considered. After all, a program that saves a high paid, senior engineer 30 minutes is worth more in terms of cost savings than one that saves a beginner draftsman 30 minutes. But if the draftsman uses the product three times a week and the engineer only once a year, then the benefits come back more to the drafting side.

Other benefits derived, such as reduced training time and improved operator morale, are almost impossible to predict with any accuracy. It should be recognized that they exist as a possible benefit when considering a customization project.

The simpler the customization the simpler the analysis in terms of cost. It is a little harder to determine the true benefits that will be derived from larger, grander customization projects. Sometimes the benefits are found outside of simple timesaving calculations. They may be found in such places as integrity of both the data and design.

Data integrity comes into play when information is passed from one application to another. An AutoCAD construction drawing of a building may contain detailed information regarding the components used to create the building. These components can be extracted into a bill of materials that in turn can be used to order components or determine the pricing. Not only is time saved by not having to re-enter the data in a bill of materials computation, but potential human data entry errors have also been averted. It's difficult to place a true value on this benefit.

Design integrity involves programs that perform some design function. If the program has been properly tested, then the resulting design from the program can be trusted to be valid. To aid in the testing of more advanced programming problems, verification output can be included for someone to check the work of the computer. When a good design program is used, the output can be used right away and the designer can move on to the next problem step.

THE POTENTIAL COSTS OF CUSTOMIZATION

Do you do it yourself, hire a consultant, or purchase off-the-shelf products? These are the primary concerns when it comes to customization of the base AutoCAD product.

More often than not, a search for a particular customization as an off-the-shelf product will yield either nothing even close to what you think you might want or something that looks like it might be interesting. On some occasions you will luck into product offerings that get you started in the right direction and leave the remainder of the customization for you to finish or contract to have finished. Don't expect to find a complete, 100% solution for your business. The better products still leave room for expansion.

When looking at a large amount of customization for a big project or something that drastically alters the AutoCAD system, the best place to start is the third party solution market. If a product does exist that is close to what you need, then it should be purchased for evaluation. With over 6000 applications written to date there is a chance you may find exactly what you are looking for. Even if the software turns out to not be exactly what you need, lots can be learned from the example. It presents an approach to solving the problem at hand. Should that approach be unworkable for your application environment, you will have learned so without a lot of effort put into customization programming.

It may not be feasible to purchase example software systems for smaller projects. In some cases, the cost of the software may be too much to consider for this purpose.

However, more often than not, the cost of the software will be low enough to justify obtaining a copy in order to learn more about the application.

There is no question that customization will most likely cost more in terms of actual money spent at first than purchasing an off-the-shelf solution. The customization begins to show a return on investment when you are able to either do something your competition cannot do or when you can do it in a fraction of their time and cost. Another area where customization is actually more frugal is when you are spending too much time and effort in searching for and then evaluating off-the-shelf solutions.

The actual costs of customization include the costs of the programming group required, be it one or many, the cost of the tools needed (more on this later), and the time. The programming group will consist of not only programmers, but also the specialists who can teach the programmer what the application is all about and how certain problems need to be solved.

Figuring the actual cost of the customization is often the goal of a good systems analysis or feasibility study. Once the tasks to be done have been broken down to individual components, the actual programming costs can be deduced and then added up to form a total project cost.

Like building a new highway, there are many steps to building a customized computer application.

HYPOTHETICAL CASE STUDY: HOW CUSTOMIZATION CAN YIELD BENEFITS

Waggoneer's Wheel Manufacturing Company produces wagon wheels for horse or ox drawn wagons like those used in the 1800s. The wagon wheels are made of wood spokes with metal strips nailed to a wooden rim. The number of spokes required for a wagon wheel, as well as the size of the wheel and axle hub, is dependent on the size of the wagon and the load to be carried. (See Figure 1.1)

At the simplest level of customization, a library of standard wheels can be created. When a customer brings in a wagon to be repaired, a trained specialist analyzes the suspension and a wheel is selected from the library. The detail of the wheel is then plotted out and a new wheel manufactured. If changes to the available wheel drawing are required, they can be made on the drawing that is subsequently saved for later usage. The amount of time saved through this basic form of customization will increase as the library grows. The problem becomes one of keeping track of the library and training sales specialists to use the library effectively.

One reason to use a CAD database to store the wheel information is that associated information can also be made available. Using the programming tools inside the CAD environment, a reporting system can be created which accesses the information

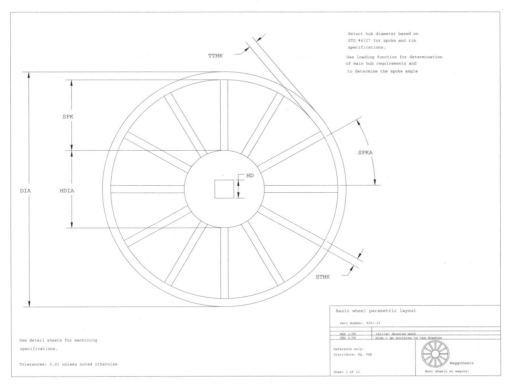

Figure 1.1 *Wagon Wheel Design*

specific to the application with a goal to then generate the bill of materials and manufacturing information. For the wagon wheels, it would be a report detailing the length of metal needed, how many nails are required, how large a shaft hole is needed, how many spokes at what size are required, how large to make the rim and so forth. The productivity gain of having the downstream reports automatically generated would be significant as designs are re-engineered.

New drawings would slow down the overall productivity of the system, making it unacceptable for use as a direct customer interface. That is, you would not want to show the customer the ease with which new wheels can be selected only to then show them that their particular wheel is not available. In this case, a parametric system may be desired.

WHAT'S A PARAMETRIC SYSTEM?

A parametric system is a graphic definition that is based on parameters. In a parametric system, the shape is defined with variables for the parameters. As the parameters change, the size of the shape changes as well.

A parametric system is generally recommended when a large number of possible permutations may exist for an item. Parametric systems create multiple variations of a design as defined by the parameters involved. As a simple example of a parametric system, consider a rectangle that has two variables: the length and the width. By supplying these values, a program can then draw a rectangle of any size in the CAD system as shown in Figure 1.2. Of course, in real world applications, parametric systems tend to be a lot more complex and can require a lot of time to program.

In the world of engineering design, parametric concepts are often employed to reduce the number of drawings required to represent a given product line. These drawings are called tabular drawings where a table appears in the drawing. The table contains variable data that is related by a name to a dimension in the drawing itself. Instead of the drawing dimension being called out specifically as is normally the case in a precision engineering drawing, the dimensional annotation contains only a variable name such as "BB" or something like that. When reading the drawing for a particular part, the table is referenced to find the exact dimension for each variable.

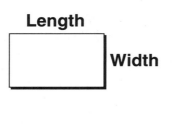

	Length	Width
A	1.5	2.5
B	2.5	3.5
C	5.0	4.0
D	7.5	4.5
E	12.0	5.0

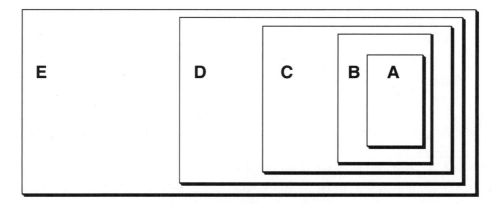

Figure 1.2 *Parametric Rectangles*

Parametric drawings save disk space in the same manner as tabular drawings save drawer space in the blue print room. In today's market of large-capacity disk drives, this is not as much of a concern as it was historically.

HYPOTHETICAL CASE STUDY CONTINUES

Getting back to Waggoneer's Wheel Manufacturing Company wheel design problem, a parametric system can also be used each time a customer specification is entered. Parameters might include information such as the diameter of the wheel, the load requirements, and the hub style. From these input parameters, a new design can be created from the parametric knowledge bank of the application. A parametric system might use a library of symbols or generate the primitive graphics directly for a wheel. As the graphics are created, any data needed for downstream integration can also be incorporated into the drawing automatically.

The primary advantage of a parametric system is that one does not have to draw each and every detail and then store it in the CAD system. Instead, we teach the system how to draw a particular type of detail and from that point on, it is capable of creating new details without the aid of a comprehensive library.

Another potential feature of a parametric system is that the users can have the choice of searching a library of predefined sets of parameters. The parameters could be stored using a part number indexing scheme to facilitate the search. Such a search system would be far superior to looking through lists of drawings.

Parametric systems can also help reduce training time. It is simpler to train a novice to fill out an input form (or dialog box) than it is to train him in the art of drawing with AutoCAD. In order to make such a claim, a parametric system would have to be capable of completing a finished drawing by itself. Such systems are rare and more expensive to create than parametric systems that only draw details. However, under the right circumstances, they may be the best way to proceed.

THE QUEST FOR AUTOMATION

When looking for CAD/CAM/CAE automation tasks that will result in significant timesaving, analyze the typical workload of the CAD operators. If the operators spend a lot of time doing something in particular, then that task may be a good candidate for automation. For example, suppose most of the time spent in preparing a drawing comes from drawing details. A library of standard details might result in significant amounts of timesaving. Taking that example further, should the detail information be selected by another program and subsequently inserted into a drawing, the timesaving can be impacted even more.

The real challenge appears when looking for automation that will significantly impact production and produce savings in the design and drafting task. Depending on whom you ask, you will get different feedback as to the best place to automate.

- Ask the CAD operator and you will find he would like to see dimensioning improved.

- Ask a designer and you will find she would like to streamline the task of passing information from her idea to the drafting system.

- Ask a CAD manager and you will find she wants a better way to track drawings and look up past work.

- Ask the plant manager and you will find he feels that improving communication between engineering and manufacturing would be best.

When beginning, all of these inputs can result in chaos for the automation expert. It's simply a matter of discovering where the highest costs are involved and seeing if they can be reduced and if so, to what extent and at what cost. This analysis will provide guidance to the automation expert so that the greatest impact to operations can be implemented.

SEEK 80-20 RELATIONS

Given the options for automation, look for 80-20 relationships. An 80-20 relationship is that which pertains to the amount of effort being employed to achieve a goal.

There are several ways to look at 80-20 relationships. The first considers that 80 percent of the work yields only 20 percent of the end product. Another view is that 80 percent of the time is spent on a given item to get 20 percent completed and 20 percent of the time is used to get the remaining 80 percent finished. In terms of customization, one area we are looking for are places where we can reduce the time required to get work done by 80 percent. Another way to look at this is to add a tool that takes 20 percent of the total time to get 80 percent of the work done.

An 80-20 relationship will generally yield a good return on investment for most customization work and should simply be considered as a rule of thumb. One will not always find such perfect relationships existing in the real world. The idea behind the 80-20 relationship is that it serves as a guide post marking where serious efforts should be considered.

Watching a CAD operator complete a drawing will often show where the 80-20 relationships exist. If they are spending most of the time in dimensioning, then a dimensioning tool could result in reducing the time requirements.

Perhaps the greatest advantage of 80-20 relationship driven automation is that the cost savings can be accounted for using the same analysis. Let's look at a basic example to help illustrate this concept.

In drawing wagon wheels, it has been observed that the operators spent time on the following tasks.

Task Description	Time Average	Percentage
Sheet layout	10	6.25
Title section draw	15	9.375
Draw the wheel	30	18.75
Dimension the wheel	30	18.75
Draw details for spoke, hub, rim	45	28.125
Dimension details	30	18.75
	160	100

Table 1.1 *Total Time to Complete Drawing*

In looking at this chart, it doesn't appear that any 80-20 relationships exist unless we combine several tasks. That is because this is not where they will be found. Instead, the tasks must be further subdivided into exactly what is going on to find the 80-20 relationships. We'll do that for one of the tasks.

Task Description	Time Average	Percentage
Dimension the wheel	30	100
- Draw Angular dimensions	5	16.67
- Draw Radial dimensions	5	16.67
- Draw Fit dimensions	5	16.67
- Draw Tolerances	15	50.00

Table 1.2 *Time Breakdown for Dimensioning*

Now we see that most of the time in dimensioning the wheel itself is spent in drawing the tolerance dimensioning. By supplying a tool that reduces the amount of time to complete the tolerance dimensions we are providing the best possible time-savings for that task. Suppose we could reduce the time required to create tolerance symbols to being just 3 minutes, which would be 20 percent of the original time of 15 minutes. The total time to dimension the wheel is then reduced to 18 minutes or 60 percent of the original time required for that step. Translating that back up to the total drawing, we find that savings of about 8% off the total will be found in just automating that one task.

The problem with looking only at the operations of the CAD system is that it's easy to get stuck in the idea that reducing the menial work will result in timesaving. Elimination of menial alone is often not enough to justify extensive customization. Many times this type of work can be reduced by using good document management and in many other cases, a good library of already-drawn components. Tasks that are considered time-consuming and menial are often good customization goals, but sometimes the cost will not justify anything too fancy. Providing a semi-automatic tool is often the best choice, as the cost of providing a fully automatic one may be too cost prohibitive.

INTEGRATION CAN SAVE MONEY

Don't overlook integration as a potential area for automation. Sometimes a broader perspective is required to see where integration can result in true savings. Even at that, it may be a gut instinct call at first. Over the years, I have discovered that integration can impact a business's bottom line more than any other form of customization.

To illustrate, consider once again the wheel-manufacturing problem. How much of an impact would the engineering information transferred directly to the manufacturing processes make on the business?

Suppose we have access to numerically controlled (NC) machine tools. The geometry information contained in an engineering document can be made to provide direct input to the NC machines. If the current practice were to manually program the NC machines from the hard copy engineering documents, the time required to get a good NC program written will vary tremendously. When a previous project can be borrowed from with minor modifications it will be possible to generate a new NC program in a few minutes. But if there are none like it, then several hours and even days may be needed—depending on the complexity of the part to be machined and the experience of the programmer—to create a new NC program.

An automatic NC program generator, on the other hand, might be able to solve the same program in just a few seconds or minutes. Each and every time it's presented with a problem of that type, the response time is now predictable.

The impact at the NC programming end is obviously tremendous, but what has happened to the engineering side? There are several possible side effects that could thwart the success of this integration because they increase the work for the engineering department.

Engineering must now create accurate documents that are consistent. It would no longer be considered acceptable to simply change the text of a dimension and plot. Instead, the geometry must be updated to reflect the change as well.

Another area of concern is that now engineering is responsible for the accuracy of the drawing and design. In the ideal world, that's how it should be. In reality, that is not how it always turns out to be. In many places it is accepted practice that manufacturing reviews the information from engineering and makes corrections as needed. For instance, corrections may be required because tooling doesn't exist in the exact sizes as called for by the engineer's design. A classic example would be calling for a 0.025" radius when the smallest tool available is 0.032" in size. The accountability issue has caused more than one integration effort to fail and should be considered at a high management level when designing a new, integrated system.

When the workload is increased in engineering in order to make the overall company operate better, some other changes will be needed. The argument that the company does better in the long run will not be heeded unless resources are made available to help meet the new responsibilities in engineering.

The important item to keep in mind is that the company can save a lot of time when an integrated system is used. The automatic wheel-making tools may be programmed in just a few minutes after a new wagon wheel is designed and accepted. Now, the only delay is in actually getting material to be cut on the machines and any assembly that may be required afterwards.

AUTOCAD PROGRAMMING IDEOLOGIES

There are several conceptual levels to consider when customizing AutoCAD: decisions must be made to run programs inside or outside of AutoCAD, with or without a prepared drawing.

The tasks that run entirely inside of AutoCAD and have no interest in anything other than the user input make up one type of ideology. These tasks include parameter-driven systems for drawing graphic and other aids to the operators. Any of the programming tools of AutoCAD can be used to accomplish various results in this realm. Obviously the more advanced programming tools such as AutoLISP®, VBA, and ObjectARX™ can provide more advanced solutions as compared to just building a menu and creating blocks.

Another ideology of customization involves using data from other programs such as word processors or spreadsheets or a design-specific application. The tools involved in programming interfaces include AutoLISP for simple text-based files, and VBA and

ObjectARX for more complicated files and for speaking to other programs through the Windows interfaces. Tasks that integrate data from multiple sources must be carefully managed in a way that insures the credibility and accuracy of the information. Passing data between programs is not a difficult task to implement, but making sure the data is accurate can be difficult sometimes.

A third ideology is to run tasks entirely outside of AutoCAD using information in the form of an export file. These types of programs do not involve graphics for the most part except as a final output or as initial input only. AutoCAD DXF™ and attribute extract files serve as useful tools in this form of customization and integration.

The most powerful customization concept in AutoCAD is the use of objects, and the best tool for manipulating objects in AutoCAD is ObjectARX. ObjectARX runs inside and outside of AutoCAD. It can access any type of file, support automation interfaces with other software systems, and use all of the possible interface tools in AutoCAD as well as in the host operating system. ObjectARX also interfaces with the AutoCAD user at a level so close as to be indistinguishable from the base AutoCAD itself. Of course, to get this level of power requires you to use more sophisticated development methods.

ObjectARX is the most complex tool of all the customization options. With diligence, it can be mastered, and once mastered, its use provides many productive roads to follow.

ObjectARX programs are implemented using the C++ programming language and a set of libraries that define a variety of classes. ObjectARX uses object-oriented programming concepts to provide a powerful interface into AutoCAD for programmers.

Because ObjectARX permits the programmer to have access to the system at a very close level, a better quality, customized computer program can be produced. User input and activity can be closely monitored so that associative data is properly maintained. The programming required to obtain this level of control is substantially more than may be required by AutoLISP or another of the AutoCAD programming options. What is reduced, though, is the potential for errors in the system as well as the ongoing support required to locate errors and problems, and then find solutions.

Determining which tool to use depends mainly on the availability of resources. It's not difficult to learn programming languages such as C++ and AutoLISP. Mastering them can take some time though. The use of consultants and free lance programmers should be seriously considered. A good consultant can save a company a lot of time and expense, as well as provide in-house specialists great examples for programming and customization of specialized tools such as AutoCAD.

SUMMARY

In this chapter we explored the various reasons as to why one would want to customize AutoCAD. The primary goals are to save time and improve accuracy. By training AutoCAD we make it work better for the types of problems we solve on a regular basis. Justifying the cost of the customization is not difficult if the impact on productivity is measured. Looking for 80-20 relationships showed how to look for the best return on investment. Finally we introduced the programming tools available for AutoCAD. The most important of which are ObjectARX, Visual BASIC®, and AutoLISP.

Before we can go any deeper into the AutoCAD customization tools supplied and the concepts behind using them, we need to gain an understanding about object-oriented programming. The next chapters introduce objects and how they fit into the concepts of CAD/CAM/CAE applications development.

CHAPTER 2

Object-Oriented Programming Primer

Although it seems like a new trend, object-oriented programming has been around for some time now. This chapter discusses the basic philosophies behind object-oriented programming and how object-oriented programming systems are structured. Object-oriented programming is not difficult to understand, and once understood, it is difficult to imagine programming without such tools at one's disposal.

COMPUTER PROGRAMMING LANGUAGE DEVELOPMENT

Computer programming languages currently employed to solve real world problems are mostly procedural languages. You define a process in an orderly fashion and the computer follows your instructions. Over the past 40 years computer languages have evolved from being very difficult to write and read—using ones and zeroes—to mnemonic assembler, where letter combinations replaced the ones and zeroes, through several generations to the modern programming languages in use today. Procedural languages in active use include PASCAL, C, FORTRAN, BASIC, and COBOL.

Virtually all of these languages support standard data representations such as numbers and strings of characters. In programmer language, they are called variables since one can place variable data into them for manipulation by the computer program. Using these data elements, powerful programs can be written to solve complex mathematical and data retrieval problems provided there is sufficient time available.

Most programmers use libraries full of useful functions. They may also resort to programming languages that are optimized to solve particular classes of problems. These approaches are legitimate for the short-term need to write programs quickly but may fall short in terms of the long-range plan. Some of the questions that have come up in the past include the following: Will the function library port to another machine architecture? Will the optimized programming language continue to be supported or even be available in the future? What if I need some data type that is not part of the standard set available? How do I make the data in my program available to other programs in the computer? How do I keep that data safe if other programs can access it?

These questions have prompted another way of looking at the programming. This approach combines all of the programming elements of the past and simply twists our perspective slightly. The approach is called object-oriented programming.

Object-oriented programming systems are typically implemented using an extended language. That is, an existing computer programming language is extended to include the features required for object-oriented programming. One of the most popular languages used is an extension of C called C++ (pronounced "C Plus Plus"). This language is what ObjectARX applications developers use when creating AutoCAD customizations.

PROGRAMMING SYSTEMS

Programming languages vary quite a bit. Ask a programmer which is best and he will undoubtedly state that it's whichever one he happens to be using currently.

Most notable of the differences between languages is the number of lines of code required to generate what is desired in an application. Assembler code requires many short lines of code to perform even the simplest tasks while extendible languages like LISP are able to do amazing feats with just a few lines of code. In some cases, object-oriented programming helps reduce the number of lines of code required to write an application. In the case of ObjectARX, fewer lines of code are generally required when compared to using the ADS library. However, more lines of code are required when compared to AutoLISP-based programming.

Other characteristics that differ between computer programming languages include the speed of operation of the final product, the ability to interface with another part of the computer system, and the way the instructions are presented to the computer from the operator. Looking at AutoCAD, we see these differences clearly between AutoLISP, ADS library-based programming and ObjectARX. AutoLISP is slower to execute and does not interface with other programs or the operating system directly. ADS library-based programming is faster than AutoLISP, yet slower than ObjectARX. Both allow direct interfacing with the operating system as well as other programs in the computer. Only ObjectARX supports advanced interfaces with AutoCAD, such as monitoring the drawing for changes in real time.

THREE FEATURES OF OBJECT-ORIENTED PROGRAMMING

Any programming language can become a tool for use in an object-oriented programming system if it meets the basic requirements. The programming language C was extended into C++ for that very reason.

There are three features that are common to all true object-oriented programming systems.

- **Encapsulation:** The ability to support objects that encapsulate both data and methods. Objects may also include other objects as well.

- **Inheritance:** Support for inheritance hierarchies where an object can share the properties of a parent object.

- **Polymorphism:** A messaging system supporting the use of common names to communicate with the various but related object types.

These features sound complicated, but they really are simple concepts. The remainder of this chapter will greatly expand on these short descriptions.

ENCAPSULATION: OBJECTS CONTAIN BOTH DATA AND METHODS

Buzzwords:

Object—a software collection of related data, methods, and other objects that describes something

Data—information about an object such as the size or location

Methods—functions that manipulate the data in an object or use the data to do something

The concepts behind objects were actually discovered early in computer science and are not all that new today. They were first invented to help simulate real world concepts and events. Initial development of object-oriented programming tactics is credited to the Simula language developed in the late 1960s. As the name implies, Simula was created to simulate things and object programming was a natural by-product.

WHAT IS AN OBJECT?

An object is a collection of both data and processes that describe how something works, how it reacts to various stimuli, and what it knows. Consequently, an object contains both data and procedures (called methods) that work with the data. (See Figure 2.1)

An object can be both referenced and manipulated. You can use the data and you can use the functions. For the most part, the functions associated with an object only manipulate the information contained inside the object, thus providing a degree of protection to that data. Often the mixture of the data with the functions is called encapsulation.

At the simplest level, an object is a collection of data logically structured to give the object some meaning to something or someone. We constantly work with objects in the real world, which is why the concept works well in simulations.

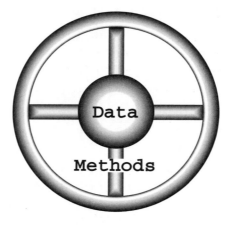

Figure 2.1 *Object Includes Data and Methods*

CONCEPTUAL OBJECT—A BUS

An example of a complex object is a bus. The data for the bus includes how many passengers it holds, the size of its gas tank, and so on. The methods, or functions, for the bus would include loading and unloading passengers, driving from one point to another, and other bus-related activities. As passengers on the bus, we must only be familiar with the loading and unloading procedures. Loading and unloading can be considered public methods. How the bus gets from one place to another is of no real concern for the passengers. That is the concern of the driver.

Driving the bus can be considered a private function or method of the bus object. There are several related activities in driving the bus such as waiting for passengers to embark or disembark as well as knowing the route to travel. And there is still more such as monitoring the performance of the engine and fuel supply.

The bus is a complex object, but as passengers we don't need to know a lot about the details of the bus in order to use it. The same can be said of object-oriented programming. There can be very complex operations at work inside the computer, but we don't need to know every detail in order to use the object.

PUBLIC AND PRIVATE

Objects can declare their elements (data, methods, and other objects) as being either public or private. When public, the elements can be accessed by functions and methods outside of the object. Public methods are sometimes called interfaces but are more commonly thought of as functions. It is through the functions that the object is manipulated. Making objects public is sometimes called exposing the object.

When private, only the methods that are part of the object can recognize, access, and manipulate the elements. Often, the data will be held private for an object so that it can only be manipulated by the methods of the object itself. This helps protect the data from outside influences.

A simple example will help clarify what an object is and how it can be manipulated. Remember that we use objects in everyday life all the time. How you look at them enables you to discover the "object qualities" that hide underneath. A real world object such as a simple alarm clock will serve as our example for showing how object-oriented programming concepts can work.

A basic alarm clock object has two data items: the current time and the time to ring the alarm. (See Figure 2.2) These data items are private and manipulated by the public methods for stopping the alarm, setting the current time, and setting the alarm time. Some private methods within the alarm clock would include functions for updating the clock display and ringing the alarm. Anything that an alarm clock requires for operation should be included in the basic object definition.

The alarm clock described thus far exists as an independent object, with its own methods for using the object. How we might choose to use it in our everyday lives differs from one application to the next.

An alarm clock object could have many applications. It could be at the bedside to wake us up in the morning. It could be attached to a safe that is opened automatically every day. It could even be tied into a chime to inform us of each hour's passing. The applications are endless.

Our alarm clock object serves as a good example in that it also demonstrates several other features of objects. Objects respond to variable events, and they limit our

Figure 2.2 *Alarm Clock Object*

visibility into things we don't have any business worrying about while still exposing the core data.

Objects can respond to variable events. The methods in the alarm clock respond to different types of events. The set time method responds to the operator wanting to reset the clock, while the alarm ring method responds to the time setting of the clock and the time setting for the alarm equaling each other. Internally there are other methods for updating the current time on a regular time interval and for displaying the time. All of these events are things that alarm clocks must know how to handle in order to be alarm clocks.

However, to use the alarm clock, we need only know how to set the current time, how to set the alarm time and how to stop the alarm from ringing.

When designed properly, objects enable the programmer to use them without fully understanding what is inside them. Consequently the objects can be changed internally when desired by the provider of the object. So long as the primary interface methods remain the same in how they accept and return information, an application accessing the alarm object will continue to run as before.

Of course, the programmer must trust that the object will behave as expected. If it doesn't, and the object comes from a reputable source such as the ObjectARX library, then the problem code should be easy to isolate. It will be located in the sections of code that access the object in question. Either some information or preparation for the object is missing, or something is not being used properly in the object.

When supplied with a software developer toolkit such as ObjectARX, object systems are generally well tested. Of course, programmers always seem to find the need for a new object that isn't in the toolkit yet, but more often than not, what is needed will be found in the existing library. Having bulletproof objects makes application programmers feel confident about the tools they are using. With a toolkit like ObjectARX, you have the assurance that even if a problem object is found, it's easy to get repairs and updates from the vendor.

Object-oriented programming promotes structured programming and structured testing strategies. In structured programming, one works with small components of code that are specific in what they do. Object-oriented programming supports this approach but allows the applications developer to think of components from a wider vantage. Instead of data being bound by simple data types, data can now be fit to the application.

Let's consider another object example dealing with engineering calculations, but without going into the math. The object is a boxed beam that carries a load. (See Figure 2.3) The data properties of the beam are the length and the thickness or width. The load capacity of the beam is based on a function that uses the thickness and shape

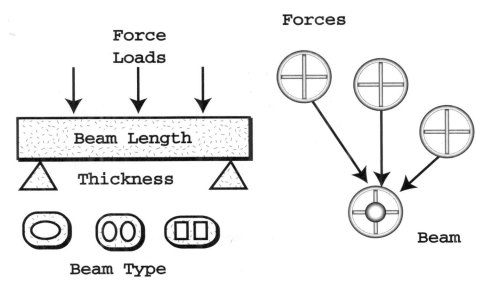

Figure 2.3 *Command Driven*

of the beam. As a beam is used in a design, the actual loads it is bearing and where it is bearing the loads is added on as additional information. The loads are computed and if the beam is strong enough, the structure stands. Otherwise, the beam is labeled as too small and another size is tried.

Different beam shapes can be added to the object so that the beam object is capable of learning more as time goes on. An interesting feature about objects is that they are dynamic in nature if we want them to be. Since most software systems are written with the idea that they will evolve over time, this aspect of object-oriented programming allows a developer to deliver modules or chunks of an application while still continuing to improve and enhance the product.

The beam object can be part of a larger structure, and as it changes, the loading on subsequent beams changes as well. With the right set of objects, an object-oriented, finite element analysis system could be created using this approach.

INHERITANCE HIERARCHIES

A major feature and advantage of objects is that they can be re-used over and over again in different ways. This saves in overall coding time and effort and is called inheritance.

An inheritance hierarchy exists when one object can be derived from another in such a way as to take on all the properties of the higher level object. Using inheritance, we can build complex alternatives to objects that are based on work already done.

When an object inherits its abilities from another object, all the features of the first object are available to the new object. New features can be added to the new object making it a unique object in itself. Time is saved by cloning the earlier object and not rewriting all the methods required to perform the basics.

Inheritance also allows objects to be modified. That is, the methods can be rewritten to accommodate some new influence or procedure. Considering the bus example again, suppose we added a new feature to the bus such as the ability to fly. The basic processes are the same from a user point of view. To use it, we need only know how to be loaded and unloaded from the bus. Getting between points just got more interesting.

The inheritance feature of objects is a very powerful programming productivity tool in that new objects can be created quickly given a good hierarchy to build from. Only the newer features need to be defined or programmed, and the new object is ready to be used. In this way, we build newer and better things from the foundations on up.

Continuing with the alarm clock example, there might be additional data and methods for an alarm clock that we would want to add (See Figure 2.4). These features might include items such as setting the alarm to a radio or a buzzer, setting the radio station, and allowing for a snooze feature. This would represent a different class of

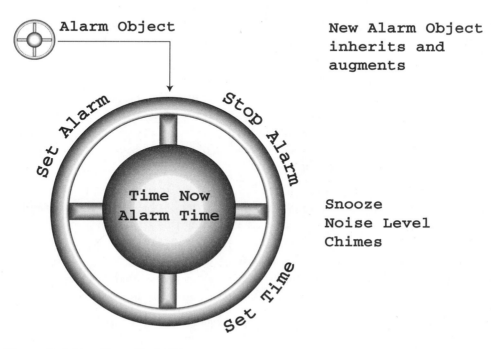

Figure 2.4 New Alarm Clock Object

alarm clock, but the basics would remain the same. There would still be a need to set the current time, the alarm time, and all the other base features of the alarm clock. In this example, the more advanced alarm clock inherits all the capabilities of the basic alarm clock and then adds some more. To create this alarm clock in an object-oriented programming environment we don't have to program any of the inherited features, only the new features.

POLYMORPHISM IN THE MESSAGING SYSTEM

The third feature of object-oriented programming languages is the use of common message names. That is not to say that the same name is used in all applications. Instead, the principle meaning is that related objects can use the same name.

One of the best examples of this concept is that of writing a program that drives various vehicles. There are many things involved in driving any sort of vehicle, be it a car, a truck, a tank, or a boat, which are common. In the English language we use the same words to describe some of these activities. For example, you steer a car, you also steer a truck, and you steer a boat. But the exact method of steering is not the same for each of these vehicles. When given the instruction to go to some point, the appropriate steering function, or method, must be called.

Using the principles of structured programming, a good programmer would have the steering functions isolated. The programmer would then use a series of tests to determine exactly which vehicle the program was driving and issue the appropriate subroutine call to get things going. As new vehicles are added to the list and the respective subroutines written, new tests must be added as well. Therefore, changes resulting from the addition of new options will require that the entire program be modified.

With an object-oriented programming system there would be no changes required to the main program when adding a new vehicle. Instead, the new vehicle would be added as a new object to the system, complete with its own data and methods. When the new object is selected for driving, the functions in the object definition are used to service the steering operation. In a C++ programming environment, the main program would have to be "re-linked" with the object library so that the new object is recognized. No coding changes would be required, just a rebuild of the program module.

PROGRAM-TO-PROGRAM COMMUNICATIONS

A messaging system is required for objects to communicate with each other. Messages are initiated by a sender and picked up by a receiver. Messaging systems have existed in various forms for years. Using traditional programming languages, messages can be relayed between different programs by writing and reading files, saving data in a common area in memory, or by using a messaging system available in the operating system (such as Windows® DDE). Of course, for this to work, all the programs that are talking to each other must use an established protocol that is the same for all members. As long as we are talking about an isolated case, such a protocol can be created and implemented by the programs involved. But when working with larger projects that are made up of numerous vendor solutions, a common platform must exist for messaging.

POLYMORPHISM

Object-oriented programming addresses the issue of a common messaging system through polymorphism. When sending a message, the target object is specified along with a particular method to be run. Additional data can be supplied as parameters to the method if so required.

A programmer might recognize this form of messaging as "calling a subroutine," however it's not quite the same. Consider that the same name can be found in a different object. As a result each method might return something different. This is where polymorphism comes into play.

The concept of polymorphism is also based on what is found in everyday life. If you ask ten people with diverse backgrounds the same question, you will most likely get ten different answers. Suppose you were from out of town and wanted to get to the airport. If you asked a cab driver, the response might be to get in the cab for the drive to the airport. If you asked someone who lived in the area, that person might direct you to local transportation with instructions on how to get there. Asking someone else who also lives in the area might result in a very different set of instructions altogether. And, should you ask directions from someone from out of town, you might get a response that says "ask a cab driver or someone else." We deal with message polymorphism all the time in the real world.

The term polymorphism is derived from the Greek and means "many forms." In programming, polymorphism means that you can use the same method to query or initiate action by an object and then get different results, based on which object you used. It all depends on which object is asked the question as to what kind of response one will get. Inside the CAD system, this concept is used with methods to draw on the screen or plotter, like the directions to the airport, and the names would be common to all graphical objects.

Using our alarm clock example once again, consider a method to set the time. Other clocks that may or may not be alarm clocks will support this same message for setting the time.

In polymorphism, the message or method name is the same, and all that differs is the object relationship. How the objects solve the individual problems is up to each object. For example, if we have two objects that are clocks, one called ALARM and another called BIG_BEN, we would want a common method name for SET_TIME.

For a programmer to set both to the same time, the SET_TIME method is called twice, first for the ALARM object and then again for the BIG_BEN object. Of course, there may be sizable differences in how each of these messages will be executed.

Simula, the original object-oriented programming language, developed the basic calling process: that is to first describe the object, then the method, and then the parameters. So a theoretical program for setting the time in both clock systems to the value in the variable "NOW" might look something like the following:

BIG_BEN.SET_TIME(NOW);

ALARM.SET_TIME(NOW);

In C++, a decimal point or dot separates the object and method. Parameters are supplied inside of parentheses (with commas between if more than parameter one is to be supplied) and the semicolon denotes the end of the logical line of program code.

One of the main advantages of message polymorphism is that changes and updates can be made to the component objects themselves without affecting the calling, or message-sending, programs. In most cases, when a new version of a tool such as ObjectARX is supplied, one simply recompiles or rebuilds the application without changing the message sending sequence. There are always exceptions, of course. New features may require additional parameters to be passed, and there may be new objects to exploit as well.

DEFINING OBJECTS—CLASSES

Objects are defined in the C++ language using what is called a "class." Essentially, a class is a template that defines the basic object qualities including the methods and data variables.

When you want to use a class in your program, you simply create an "instance" of the class and then supply your own variables. The class defines the methods. So when you want to manipulate the data using one of the methods, you reference the particular instance by name and then supply the appropriate method and parameters as needed.

When programming in an object-oriented environment, it's common to define classes and instances of classes supplied by various software modules. The most frequent approach to such programming is to use instances of modules supplied by vendors such as Autodesk, although many companies are now developing their own object systems.

Let's relate object-oriented programming technology to AutoCAD usage. Suppose your program was to create a new line in the current drawing. In object-oriented programming terminology, a new instance of the line object needs to be created and then the method to add a new line to the drawing is invoked to create the new entity. This may sound complicated, but it is really easy to implement using languages such as C++ and the ObjectARX library.

OBJECTS ARE STORED IN LIBRARIES

Objects are stored in libraries of classes. When you reference a single object, it may in turn reference other objects. Just as a book taken from the library often contains bibliographies and references to other books, objects may reference components from other objects. Thus a library is a good way to store all that information. The flip side to this is that the entire library will be mapped into your application, even if you don't use very much of it. In the case of AutoCAD, there are several libraries supplied for each of the logical components of the system. For example, there is a library for geometry manipulation and another for database components. These libraries all contain multiple class definitions and, in the majority of cases, an ObjectARX program will contain all of them.

Conventional programmers should be cautious about the term "library" as it is used in this context. For a conventional programming system, a library contains only subroutine modules that can be called. The link program module obtains only those components desired and maps them into the application program. A library in the context of an object-oriented programming system contains the class definitions that in turn contain the methods as well as templates for the associated variable data.

In ObjectARX, the library a particular class is located in can be determined by looking at the first couple of characters in the name of the object. For example, anything starting with the characters AcRx is from the run time class registration set while objects starting with the letters AcDb reference the database classes. These library names are a bit daunting at first, but with a little practice it becomes easy to find what you need.

Object-oriented programming works inside the computer by employing a concept that is called late binding. When your C++ program is created by a compile and link operation, it references the object libraries. At that time, two kinds of bindings or links are created. In computer program parlance they are called early- and late-binding functions.

The first kind is a direct linkage to a subroutine that is mapped into your program file. It is called early binding because the linkage is established when the program is built, and a copy of the subroutine is placed into the program. This is the more traditional approach used in writing programs.

The second approach is when a pointer name is established to an object library reference. This is late binding since the actual linkage does not take place until the program has started to run. Late binding makes object-oriented programming possible. When a late binding is requested, the module (or modules) responds if it is loaded into memory. If the module is not in memory, the system will attempt to locate and then establish communications with them. Failing to find the modules means that the objects are not available for that computer system, and your application must decide what to do in that case.

HOW DOES OBJECTARX WORK?

ObjectARX is a set of libraries defining the AutoCAD objects that are public. When your C++ application program is created, you must include references to these libraries. The application program you create is actually a dynamic link library or DLL. As the name implies, the library is dynamic in that it rolls into memory when requested. Most of the time, the file extension for ObjectARX programs is .ARX instead of .DLL, even though they are the same in the eyes of the operating system.

AutoCAD starts your ObjectARX program by loading it. The load is initiated either at the start of AutoCAD, when first requested, or by the AutoCAD command ARXLOAD. Most of the time, ObjectARX modules are loaded automatically when being used at a given workstation. To load the modules automatically one simply includes the name of the program to load in a text file list.

After loading the ObjectARX module, AutoCAD calls the primary entry point subroutine to establish communications and give your program a chance to define all of its requirements. The primary entry point function has the same name in all ObjectARX programs. AutoCAD expects to find the entry point function in order to successfully integrate the application into the AutoCAD workspace. The primary entry point function has a fixed parameter list as well. The parameter list contains a message code that is one of a set of values. These values tell your program what is expected of it.

One message to the primary entry point function is to initialize your program. This is sent when the program is first loaded. At this time the ObjectARX program can declare new command names and AutoLISP function names if needed. It can also define which objects of AutoCAD it wants to have direct communications with and establish pointers between the associated objects. The pointers established are the subroutines in your application that will be called directly when something happens. These subroutines then become additional entry points into your program from AutoCAD.

Another type of message is a code number requesting the execution of a function defined to AutoLISP. ObjectARX programs can define new AutoLISP functions for use by applications in that language. When a function is defined, it's given a message code number by your program. This message code number is sent from AutoLISP when a program is running that needs your function to return a result. Both ADS and ObjectARX support the extension of AutoLISP in this manner, thus enabling the developer to use all the tools to the best of their ability.

Additional message codes inform of new drawings being loaded, drawings being saved, and requests to unload the ObjectARX application. The primary entry point program can be thought of as the main program section, since the start up and unload requests are sent to this function by AutoCAD.

WHAT'S SO SPECIAL ABOUT OBJECT-ORIENTED PROGRAMMING AND AUTOCAD?

The answer is found both at Autodesk, as well as with application developers worldwide. Depending on one's perspective, there are numerous advantages to Autodesk having taken this direction with the customization of the CAD system.

First and foremost on the list is the ability to add new objects by Autodesk or anyone else who needs to do so. Consider the AutoCAD enhancements for Autodesk's software product, Mechanical Desktop®, which include new object definitions for the advanced 3D modeling that is found in the package. The new 3D objects integrate directly into the AutoCAD structure, allowing for further extension by other developers. The addition of new objects is of primary interest to Autodesk and serious application developers. There is a substantial amount of code involved in creating a new object of any significance, but by using inheritance of the objects a lot of the headaches are diminished.

Of primary interest to application developers is speed of execution. The direct linkage with routines in memory delivers the fastest possible response times between your application modules and AutoCAD. In fact, your programs access the AutoCAD utilities at the same level as AutoCAD. It just cannot get any faster.

Other developers are intrigued mostly by the fact that ObjectARX opens doors for direct contact with other areas of AutoCAD. These doors have been closed off in the past and are not available via AutoLISP or by using just the ADS library. Having the ability to be notified when specific events are taking place is a powerful tool. Suppose you have an application that is concerned about the integrity of objects in the drawing database. An ObjectARX routine can be written that requests a response each time the drawing database is updated. Although your routine cannot stop the activity from taking place, it can take whatever other countermeasures may be required.

Another feature of ObjectARX is that coding requirements are reduced when compared to ADS-based programming. This fact is particularly evident when working with entities. Under the ADS programming model, one works with result buffers that are chained memory items. It is the program's responsibility to loop through the result buffer chain to find the specific information it requires, such as a layer name. In ObjectARX, there is a method for each and every object in the drawing database to get and set data elements such as the layer name.

By using object-oriented programming thinking, ObjectARX gives application developers the opportunity to expand AutoCAD into specific application areas, and still retain the basic system functionality. As seen in some of the more advanced ObjectARX applications such as Mechanical Desktop, AutoCAD can be morphed into a powerful 3D modeling system. Yet the interface is clearly AutoCAD and the basic abilities still exist.

If integration is an issue, then ObjectARX also provides one of the best solutions. Object level information can be shared with other applications by making the objects public. If you need to transfer data between AutoCAD and another application, an object-oriented programming approach should be considered as a primary alternative.

The key items that object-oriented programming brings to the AutoCAD developer are a very complete set of routines for accessing and manipulating drawing data as well as a set of tools for monitoring the system at a primitive level. Using these features and the concepts of object-oriented programming, one can create very advanced enhancements that go far beyond basic menu macros and libraries of shapes. The use of objects, real-time speed, and control features of the ObjectARX library makes it a tool that has no parallel when customizing AutoCAD.

Another aspect of ObjectARX is that it points in the direction Autodesk is taking for further developments of the CAD system. As a system manager the ObjectARX approach to adding new features is wonderful since the basic AutoCAD system remains untouched as new modules are tested and brought online. And should a new release of AutoCAD be shipped, one only has to rebuild the application to establish the linkage information for late binding in the newer modules. AutoCAD enhancements such as Mechanical Desktop and Visual LISP™ demonstrate this feature clearly. Should you need to use the basic AutoCAD only, these features can be turned off and AutoCAD is restored to "normal." The modules are not in memory, do not interfere with AutoCAD, yet remain easily accessible when needed.

SUMMARY

This chapter introduced objects and how they came to be in a historical sense. The evolution of programming languages into object-oriented systems has greatly changed the way new systems are developed. Objects have three basic features in the form of encapsulation, inheritance, and polymorphism. That is, objects are self-contained software packages inside the computer, they can use and reference each other, and common things share common ways of programming them.

Objects are implemented as dynamic link libraries (DLL) in the computer with some aspects of the objects exposed as public and others as private. The AutoCAD ObjectARX tools allow programs to be developed that link up with AutoCAD via a series of messages and can then become new commands or features of AutoCAD. Through ObjectARX new objects can be added to the CAD/CAM/CAE environment.

CHAPTER 3

Evolution to Object-Oriented CAD/CAM/CAE Systems

Object-oriented CAD/CAM/CAE did not just happen overnight. Learning about the development that lead to today's design tools provides a path to understanding the real power behind the concepts. This chapter explores the past developments of CAD/CAM/CAE technology and looks at how they relate to object-oriented programming.

AUTODESK OBJECT-ORIENTED DIRECTIONS

The historical evolution of CAD/CAM/CAE systems in the past quarter century sets the stage for understanding why object-oriented programming works well in the context of such systems. A brief history of earlier AutoCAD drawing storage and structure is used to illustrate how object-oriented thinking represents a quantum jump forward. Autodesk is one of the leaders in pursuing this improved way of customization with the ObjectARX tool kit.

A CONSTANT PARADIGM SHIFT

"Paradigm shift" is a term that is used a lot these days. Paradigm means a representative standard pattern or form. It's an accepted way of thinking or doing whatever is being discussed. Thus, a paradigm shift is when the standard pattern or concepts behind something change.

Computer systems, hardware, and software are in a constant paradigm shift. At least, that's how it appears in the CAD/CAM/CAE market. Vendors are quick to announce new paradigm shifts in technology, and in the majority of cases they are right in making such a claim. More often though, the shift is so small compared with neighboring technology that it's lost on customers who may not understand where the technology was and what they are seeing. The underlying reality is that there is a constant shift underway as new developments take place.

Hardware and software have followed their own paths of development. In most cases, the hardware is ahead of the software. That is to say that hardware must exist before software can be written to run on it. Sometimes software is created that needs more muscle power. In that case, software gets ahead of hardware. The two are deeply intertwined in the history of computer technology and pull at each other to race ahead.

MICROCOMPUTER EVOLUTION

In the early 1980s, very few people envisioned the power of the microcomputer and the speed that would be attained. There were some visionaries that recognized the microcomputer for what it was, but most computer scientists considered it to be a miniature version of the computer as they knew it—a smart terminal at best. Generally, it was believed that the microcomputer would never rival the power of a centralized, large computer system. After all, as that technology improved it would be absorbed into the central systems. It was ludicrous to imagine that desktop and portable computers would be capable of running circles around the mainframe and super minicomputer systems of those days. It's just as hard for us today to anticipate to what level palm top computers will evolve and what user interfaces will look like in the near future.

During the 1980s and early 1990s, rapid performance increases in microprocessor technology took place. Early microcomputer systems ran at clock speeds of two to four megahertz. Microcomputers today run at a hundred, and more, times that speed.

At the same time that microprocessor speeds have increased, the cost for computer memory (RAM) has decreased significantly. Initial microcomputer systems only had a couple kilobytes—thousands of bytes of storage—of main or high-speed memory. By the early 1980s a popular computer system based on a microcomputer had between 64 and 640 kilobytes of high-speed memory. By the early 1990s, main computer memory was being measured in terms of megabytes—millions of bytes of data storage. And by the time Windows® 95 became popular, it was not uncommon to find computers with 32 or more megabytes of main memory. This is an amazing situation when one considers that just fifteen years previous a large and expensive disk system for a microcomputer had 32 megabytes.

Disk drive memory has also decreased in cost while at the same time disk drives have increased in capacity to staggering levels. As just mentioned, a large disk system in the early 1980s was about 50 megabytes. In the mid-to-late 1990s, disk drive systems are measured in gigabytes; billions of bytes of storage, and are a fraction of the cost of the total computer system.

All this has led to computers sitting on desktops that are capable of advanced graphics, advanced mathematical modeling, and publishing. As a result, microcomputer applications in CAD/CAE/CAE have also seen tremendous growth.

MICROCOMPUTERS IN CAD/CAM/CAE

The first microcomputers to be applied to CAD/CAM/CAE systems were in computer graphic display systems. Maintaining the cursor position and displaying it on a high-resolution screen was the first CAD/CAM/CAE application of microcomputers in the 1970s. As the available power in microcomputers began to increase, so did the processing tasks undertaken by the microcomputer. It was only a matter of time before someone condensed the functionality of the mainframe and minicomputer CAD/CAM/CAE software into a commercial microcomputer system.

In the early 1980s the first CAD systems were introduced for commercial microcomputer systems. AutoCAD was one of them. Several other companies also began marketing their microcomputer and desktop minicomputer based CAD/CAM/CAE solutions during that time frame as well. The cost of these systems varied tremendously as did the marketing strategies of the various companies involved.

Some of the software vendors started selling their CAD/CAM/CAE solutions for microcomputers through established sales forces. Larger companies who had already been in the business for several years touted their microcomputer solutions as being low-end workstations for casual users. Others took their products right to the streets and sold them through computer outlet stores called dealers.

The initial drawing software for microcomputer systems was not comparable to the mainframe and minicomputer systems that were commercially available. There were many features found on the larger system that microcomputer-based CAD did not offer. But they did show a lot of promise, even though it was hard to imagine a microcomputer of the power available today. At that time, even large super minicomputers did not sport the capability one can put on a desktop currently.

DISTRIBUTED COMPUTER SYSTEMS

The idea of a distributed computer system, where each user has a computer to privately use and abuse, is not a new one. It has simply been a matter of two technologies catching up to the price the market was willing to pay. Those technologies are microcomputer systems and networking structures. Microcomputer systems needed to improve the power delivered, and networks needed to become faster, more reliable, and more inexpensive. Both situations have already happened and are continuing to improve. Microcomputers are more powerful and fast networks are more economical.

CAD/CAM/CAE is well suited for distributed processing. Suppose you are working on a large drawing while your neighbor is crunching numbers. If the two of you shared a single computer, each process would run slower than normal because together they are making heavy use of the mathematical computation skills of the machine. Users will be quick to share how frustrating that can be. Having dedicated use of an isolated computer for CAD/CAM/CAE applications is by far the best solution.

The power delivered by microcomputer systems, and the distributed computer model, combine to provide a very strong platform for advanced computer graphics and CAD/CAM/CAE systems. Engineering power tools are a reality, at least as far as the hardware is concerned.

GRAPHIC DATA: FROM POINTS TO OBJECTS

Even though it may appear that hardware is outpacing the software, it really isn't in all application cases. CAD systems and users have been waiting for the hardware to get better.

The earliest versions of CAD did not have a lot of computer power to exploit and made up for it by reducing the amount of work required to manipulate the graphics. The CAD graphics were basically stored as points with a flag indicating whether or not to draw to the point with the pen down, and leave a line in the process, or to move to the point with the pen up, and leave no mark. Additional flags were found, along with alternative record types for defining other related information, as the systems grew more sophisticated. Drawing the CAD graphic on the screen or at the pen plotter was a matter of following the dots and leaving a mark when told, and nothing otherwise.

Obviously this strategy of storing points consumed a great deal of disk space when drawings were made that involved curves. But most two-dimensional engineering and architectural drawings contain lines, and consequently, the method worked reasonably well. Three-dimensional drawing was not considered something that engineers and architects would need, and if they did, more expensive computer graphic systems dedicated to making images could be purchased. Integration between these two systems was virtually non-existent. The mathematical objects found in the higher end graphics system required much more information than connected dots.

To give the point groups more intelligence, additional record flags were added to earlier databases. These record flags allowed points to be grouped together into a primitive object. There was a record in the database that indicated the start of a conic section. A series of points would then follow which described the conic and that would be terminated by another record indicating the end of the group. Although now the object was recognized as a conic, there was no information indicating how it had been created in the first place. The next logical step was to store the input parameters along with the points. That is where distributed computer technology resurfaces.

GEOMETRY OBJECTS

Using a dedicated computer of sufficient speed, one does not have to store all the dots or vectors. Instead, the computer graphics program can recalculate the vectors as needed, given only the parametric description of the object to be drawn. This strategy greatly reduces the amount of storage needed to house a drawing when lots of curved objects are involved. It also improves the methods of communication between two-dimensional drafting and three-dimensional models, as well as integration with other geometry-based applications.

It's interesting to note that AutoCAD stores objects at multiple levels while you work with them. Curved objects are converted to dots and vectors which are stored while the object is visible. When programming applications work with AutoCAD, you do not have to concern yourself with this fact because the objects are updated automatically whenever they are manipulated.

From the earliest versions, the data structure found inside AutoCAD was object-based. An arc object contained parameters describing the details of the arc, such as the center point, radius, starting angle, and ending angle. Conventional minicomputer-based CAD/CAM/CAE systems of the time-contained point-based databases, with support codes included, to help indicate basic geometric objects such as arcs and circles. The use of basic parameters only was a new way of thinking in CAD database implementation. Considered impractical previously, the door was opening to more advanced ways of thinking about data and the work involved to write an application program.

OBJECT-ORIENTED THINKING

Objects in CAD/CAM/CAE greatly aid the applications developer who is adding discipline specific modules to customize the system. For example, consider the basic problem of finding the intersection of two graphical entities. In a point-based system, the program will have to test all the points associated with the entities to find the intersection. In an entity-based system, the points of the entities must be calculated, and then intersections sought, after using specific routines for each type of entity.

An object-oriented programming system provides an intersection routine through polymorphism that simply accepts two objects and returns a list of the intersections found. The intersection function itself is part of the system and has been exposed to the programmer to use as desired. What happens inside the intersection routine is of no interest to the programmer so long as a valid result is returned. From an application programmer's point of view, the object-oriented programming approach is much simpler and requires fewer lines of code.

The real advantage of object-oriented programming for most CAD/CAM/CAE programmers is that the manipulation of the graphics can be done by the package that does it best. A custom program does not have to contain routines that duplicate the features already contained in the system such as finding intersections. This allows the developer to concentrate on the application at hand and not have to worry about the geometric solutions needed to achieve primitive operations.

OBJECT-ORIENTED PROGRAMMING LANGUAGES

The object-based structure of AutoCAD entities has been in place for some time. All that was needed was a strategy by which they could be made accessible to the developer community. In the mid-1990s, a method became popular by which this task could be accomplished within AutoCAD. Object-oriented programming tools had become available, enabling the transformation of entities in the database into objects. The newer objects not only contained the parameters needed to render them, but also contained the processes that allowed the objects to be manipulated.

It has taken some time to implement these strategies into the complete AutoCAD product. Subroutines that manipulated the geometry had to be adjusted to fit into the object structure, and a series of libraries defined which offered access to the objects. The result of this effort is an interface tool called ObjectARX, which permits direct access into the very heart of AutoCAD and goes way beyond the basic database interfaces. In fact, ObjectARX is one of the most advanced CAD customization tools ever provided by a leading CAD supplier and fits well with the Autodesk reputation of setting the pace in this regard.

DRIVING THE SYSTEM: COMMAND STREAMS

The customization of AutoCAD typically involves the clever combination of commands and procedures to create macros. This type of programming involves sending streams of commands to the system to process. As a result it is often called command stream customizing or programming. Command stream programs take on several different forms in the AutoCAD environment, ranging from menus which may contain several statements in a preset order, to AutoLISP functions which use the commands to accomplish more sophisticated manipulations of the graphics. Menu macros and AutoLISP provide powerful tools for accomplishing much of what someone would want to do in terms of streamlining AutoCAD to a given discipline.

The principle drawback to command-stream-based programming is error control. Command streams are a one-way communication. Command requests are made with no regard for the results. If an error occurs during the command execution, the calling program has no way of detecting the failure and will most likely continue to supply parameters until canceled. The more complex the command stream, the more likely a problem can occur along the way.

ENTITY MANIPULATIONS

Although many wonderful routines can be created using commands, more advanced AutoLISP and ADS functions may also directly access the drawing database for manipulation of the objects. These entity manipulation tools allow developers to go beyond basic command streams when needed. However, to effectively use these tools, the level of programmer knowledge required is increased, and in some cases, more programming is required to finish the task. The trade-off is that this deeper level of access into AutoCAD can be used to create tools that are less error prone.

OPERATION ERROR PROTECTION

Entity, or object level accessing programs tend to be less error prone since they are not controlled by other settings in the system. Certain conditions change the way commands work, or more specifically, change the order of the information to be supplied. The TEXT command is a good example of this sort of activity. When a Text Style has been defined with a constant text height, the TEXT command sequence changes as it will no longer request the height to be applied. Normally, the TEXT command requests the text height using the last text height as the default. If you had written a macro command sequence that expected the text height to be set to a constant, the command stream will be compromised if the current text style does not comply.

OBJECTARX OFFERS EVEN MORE

Most AutoLISP and ADS applications programmers may wonder, "How does it get any better using ObjectARX?" The answer is simple. We now have access to the same tools used by the Autodesk development team to add new features to the system. If you have ever developed applications for AutoCAD using AutoLISP and ADS, you most likely have discovered their boundaries in relation to the overall operations of AutoCAD.

ObjectARX opens new doors into AutoCAD for applications programmers. Not only is the database exposed, but other key areas of the operations are exposed as well. Programmers outside the company can access the same libraries used by the Autodesk programmers when adding new features to the system. That means that enhancements can be integrated so tightly as to be indistinguishable from AutoCAD itself.

But there is even more that is available in ObjectARX. ObjectARX provides facilities for interfacing with the events of AutoCAD on a very fluid level, thereby tightening the level of integration available. Events such as starting a new drawing, adding a new object to the drawing database, moving or removing an object, and saving the drawing, are now open for customization.

NATURAL EVOLUTION

Programming AutoCAD has evolved from simply pushing commands to actually integrating into the system entirely, so that our applications become part of AutoCAD. The expertise required to implement the more advanced tools is substantial in comparison with that needed at the basic level, however the potential returns are significant. This is especially true for developers who create programs used by many people in multiple locations. Applications can be tightly woven into the fabric of AutoCAD so that mistakes or invalid processes can be detected and dealt with before becoming a problem for the user.

SUMMARY

The evolution of CAD/CAM/CAE systems from basic drawing to advanced object manipulation has been swift. Most of the important elements of distributed processing and computer graphic data manipulation had been solved however the hardware industry needed to improve. In doing so, the software once again fell behind and had to catch up to bring today's powerful object-oriented CAD/CAM/CAE environment.

Programming CAD/CAM/CAE applications has come a long way in the past two decades. From point-to-point type to entities and then to objects is quite a journey. And there are many more exciting things on the horizon as users and programmers alike create more objects for manipulation.

Object-Oriented Programming Inside AutoCAD

ObjectARX is provided as a set of libraries that are divided up by the type of functions they offer. An understanding of the libraries and how they can be utilized is presented in this chapter which introduces the ObjectARX environment.

OBJECTARX

The ObjectARX programming system for AutoCAD contains a comprehensive set of libraries for interfacing with AutoCAD using the C++ object-oriented programming language. The ObjectARX libraries provide the tools needed to access AutoCAD at a very deep level, that is, at the same level that the Autodesk programmers write new technology for the base package. As a result, ObjectARX libraries give us access to a powerful set of functions for manipulating graphical entities. Not only is the AutoCAD database exposed, the basic operations of AutoCAD are exposed as well. Another way of looking at this situation is that as entities are manipulated, ObjectARX programs can be notified directly. This gives the applications developer complete control over the runtime CAD/CAM/CAE environment.

LIBRARY, CLASS, AND DLL

Three terms we need to have a comfortable understanding of to get much deeper into ObjectARX are: Library, Class, and DLL. They are introduced in the following paragraphs. These terms come from the C++ programming language environment as well as Windows.

When an object-oriented programming tool is provided to someone else writing in the C++ programming language, it is normally provided in the form of a library. A library is made up of several items. There is a part of the library used by C++ programs during development and another portion that is used when the program actually runs in production.

Typically there is a file (or files) containing the definitions of the functions and data types that a C++ program can utilize. These are the class definitions used in the C++ language for that particular object set. The word "class" is a keyword in C++ that is used to define the contents of an object. When several classes are grouped together, this is called a class library.

Normally the class library definitions are supplied in the form of ".h" header files. Class definitions are provided in an accessible format for reference purposes as well as for inclusion in your C++ programs. Any text editor can be used to display and print these files. However, only programmers who have experience working with them should exercise the liberty of changing them. Easier to read information about the classes can be found in the online help files and reference manual supplied with ObjectARX, but the header files are used when creating applications and as such serve as the final reference when a question comes up.

Another part of the library is delivered in the form of LIB files. Most programmers think of LIB files as being "linker libraries." LIB files contain subroutine definitions ready to be mapped into a program. They are binary files that are ready to be incorporated into a program when the link operation is performed. Linking is the last step before a program is turned into an executable file. Subroutines in a LIB file fall into two general categories. Either they fulfill a job internally by themselves, or they provide a bridge to link into an object that is elsewhere in the system.

It's important to keep in mind that the term "library" goes way beyond the LIB file alone. A library is a collection of objects that are related in some fashion. A LIB file is a library of subroutines for the link program to use when creating an executable module. A library of objects typically is made up of a LIB file as well as others.

The last part of a library is found in the computer when our application runs. Exactly where the functions will be in the computer is unknown when the program is first written. Instead, the locations are determined dynamically when the program is run at the machine.

The dynamic part of an object library is normally delivered as an executable module for the computer. It is in binary form and immediately ready to run on the machine. There are two ways to deliver the executable modules. The first is to supply a program module that is started by the user. The second, which is used in Windows programming, is to provide a DLL (dynamic link library) file that is started by another program. Additionally, there are hybrids such as AutoCAD, that provide executable and DLL files in their packages.

The primary difference between the two types of executable modules is that the user starts one, and the other is started by another program. For example, the user starts AutoCAD to create or edit drawings. When AutoCAD is running, it may start

DLL programs to handle certain situations or enhancements such as the Mechanical Desktop product, AutoCAD MAP®, or your custom program.

When the executable portion of the object library is running, any objects that these programs have defined are now available. This is how an ObjectARX program interfaces with the AutoCAD system. ObjectARX modules are DLL files and are loaded on request by AutoCAD. Once loaded, the objects in the ObjectARX program are available to AutoCAD, and the objects in AutoCAD are available to the ObjectARX module. Thus, the ObjectARX based DLL becomes a plug-in module for AutoCAD and expands its abilities.

OBJECTARX LIBRARY NAMES

ObjectARX is provided as a set of libraries, that is, sets of organized class definitions for C++ programmers. Each library serves a specific purpose and all communicate with each other as needed. The libraries have simple four letter names that begin with the letters AC (for AutoCAD) followed by two more characters designating a particular library. For example, AcDb is used to designate the AutoCAD database library.

All objects inside a given library have names that start with the same four characters, making the pedigree of an object easier to trace back to the header files. The long names that at first seem to be of random origin are in reality developed in a logical manner. For example, object named AcGePoint3d directs you to the AutoCAD geometry header section for the exact definition.

The libraries supplied with ObjectARX are as follows.

AcRx	System level routines for linking the ARX DLL program module to AutoCAD also include basic object classification tools.
AcEd	Editor services for adding new commands and for systems level event notification requests.
AcDb	Database object definitions for all geometry and table items inside AutoCAD.
AcGi	Rendering interface library for drawing objects.
AcGe	Geometry definition library for elements of objects such as points, along with tools for performing geometry data manipulations.
ADS	Utilities for selection set and entity manipulations in an interactive environment for R14. These utilities are now integrated into ObjectARX 2000's libraries listed above.

Table 4.1 *ObjectARX Library Names*

We will explore these libraries in more detail, but for now look at how ObjectARX programs merge with AutoCAD at runtime.

LINKING UP WITH AUTOCAD

The best place to begin is where the application program actually starts. It must first link up with AutoCAD when it is being loaded into memory. The linking process actually involves several steps, but is quite easy from an application programmer's point of view. Most of the difficult work is completed internally by AutoCAD when it loads up the ObjectARX module.

In a typical C++ program, execution starts with a routine called "main." In an ObjectARX program, execution starts with a routine named "acrxEntryPoint." All ObjectARX programs must have a function with this name so that AutoCAD can start up the conversation. When the ObjectARX program is loaded into memory by AutoCAD doing an ObjectARX load operation, AutoCAD will initiate communications by placing a call to the entry point subroutine in the new module.

The call to the entry point contains a message code number and a data area where AutoCAD can pass information about the application status for locking and unlocking purposes. (See Figure 4.1)

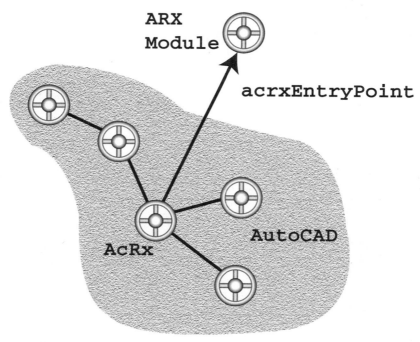

Figure 4.1 *acrxEntryPoint Starts the Linkage*

When the entry point subroutine receives the message code from AutoCAD, it is expected to act on it immediately. There are several possible requests that AutoCAD can send to an ObjectARX module through this mechanism. The most important and most frequently used are listed in the table below. There are others that may or may not be important to your application. They can be referenced in the ObjectARX Developers Guide, available from Autodesk.

Message Name	What It Means
kInitAppMsg	Initiate the communications link.
kUnloadAppMsg	Request to unload the application from memory.
KLoadDwgMsg	A new drawing has been loaded and AutoCAD is fully functional.
KUnloadDwgMsg	Drawing is being released; time to release memory and database items.
KInvkSubrMsg	A request has been made to run a subroutine registered as an extension to AutoLISP through the ads_defun() function call.
KEndMsg	The drawing edit session is ending with a save.
KPreQuit	The drawing edit session is about to quit, but your ARX modules have not been instructed to exit as yet. Good time to clean up files that may be open.
kQuitMsg	The drawing edit session is ending without a save.

Table 4.2 *Entry Point Messages*

The message names are actually integer numbers that have values associated with them in the header files for the ObjectARX library. It is strongly advised that these names be used instead of the integer equivalencies in case these values are changed at some time in the future.

The first time the entry point function is called by AutoCAD it is to initialize the communications. This may occur when AutoCAD is not completely up and running inside the computer and happens when AutoCAD is first loading and the ObjectARX module is designated as being part of the system. You cannot expect to perform commands and similar operations during this first call. You will have to wait until there is a second call informing your program that AutoCAD is ready to do anything. Besides, there are other tasks to be completed during the first run of the entry point.

Your program will be contacted when a new drawing is loaded when operating in a multiple document environment (MDE). The drawing loaded message will be sent to the entry point informing your application that the new drawing has been loaded. For most applications, this is when any required memory buffers will be allocated for your application associated with that particular drawing.

One of the more important activities that takes place when initializing the first run of the entry point subroutine is that pointers can be set up between the custom application and AutoCAD. A pointer is the value of a location in memory where a particular program's functions and data can be found. The speed of the ObjectARX interface is achieved through direct memory address pointers that allow AutoCAD to call the custom routines directly. At the same time, the custom application obtains the pointers to the AutoCAD objects to obtain similar performance abilities. (See Figure 4.2)

The second time the entry point subroutine is called is when a drawing has been loaded. Now AutoCAD is fully functional and any operation desired can be performed. The command prompt is about to be presented to the user but the ObjectARX programs get a stab at the system first.

When the drawing loaded message has been received, it's time to define any new AutoLISP functions—added via the ads_defun function—and to set up the drawing. Layers, text styles, and so forth, can now be checked and/or established in prepara-

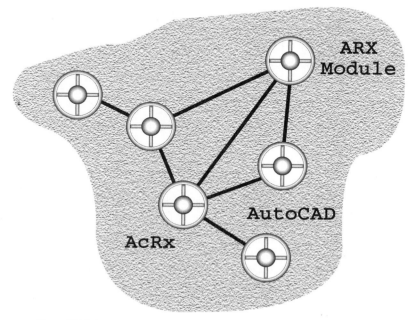

Figure 4.2 *ObjectARX Module Absorbed into AutoCAD*

tion for whatever tasks are to be solved. Since AutoCAD is fully functional, any command or manipulation can be utilized.

It is important to remember that your ObjectARX entry point module will get two messages when it is starting up. The first is to initialize new commands and other deep system level items. When the message comes in, AutoCAD may or may not be 100% functional. It's best to assume that it's not. The second call into the entry point, with the message meaning that a drawing has been loaded, signifies that AutoCAD is operational and you may initiate commands. This is also the time to define new functions for use in AutoLISP programs.

ObjectARX programs can be activated after a drawing is already loaded. In this case the messages will follow each other in logical order. First the initial contact will be made and then the drawing loaded message will be sent.

Once your ObjectARX module is running it will not receive the initial contact message again. When a new drawing is loaded or started in the editor, the drawing loaded message will be sent. Do note that a message is not sent when the operator switches drawings in the multiple document interface of AutoCAD 2000. Your application should be able to handle situations where that will arise to be 100% MDI-compliant.

ACED: AUTOCAD EDITOR LIBRARY

A library used extensively when a custom ObjectARX application is initializing is the Editor Library named AcEd. The AcEd library contains the routine that allows ObjectARX programs to declare new commands to AutoCAD. In ObjectARX, commands are added to AutoCAD by appending them to the command stack. This is the same process that Autodesk uses when setting up localized (language specific versions) copies of AutoCAD.

The command stack is an object (class) defined in the AcEd library. The command stack contains the commands available to the user during the current editor session. Through the use of the function "addCommand," a new command is added to the command stack, along with the memory location where the command is serviced. The location where the command is serviced is the subroutine address you have written for performing this new command. The exact subroutine address that handles the command being added in the custom program is determined at runtime. Programming languages such as C++ have simple procedures for obtaining such information. The address is then supplied as one of the parameters to the addCommand function so that the command processor knows whom to contact when the command is used inside the AutoCAD system.

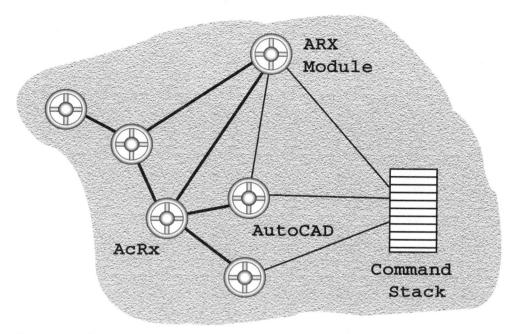

Figure 4.3 *Command Stack Pointers*

By having memory address pointers from the command stack point directly to the ObjectARX functions (See Figure 4.3), AutoCAD spends the minimum amount of time getting to the custom program. The sharing of AutoCAD memory space allows ObjectARX programs to respond and run fast, just as new enhancements to AutoCAD should.

The usage of the main command stack also means that any time the new command is used in AutoCAD, either from a menu, command line entry, AutoLISP (command) expression, or from ADS, the ObjectARX subroutine will be run. Commands created using the command stack are considered native commands inside AutoCAD.

The DLL approach to AutoCAD customization is the fastest possible interface available, and it provides more functionality for the developer as well. CAD/CAM/CAE customizing is much more than supplying commands and manipulating entities fast. Because the custom ObjectARX program resides in memory alongside AutoCAD, references to the ObjectARX program can be made as seemingly unrelated events occur inside AutoCAD. In other words, AutoCAD calls the ObjectARX functions because something has happened, not just because a command was requested. Thus, ObjectARX programs have the ability to monitor events such as the manipulation of entity objects, command streams, and general transactions inside AutoCAD. This whole approach represents the state of the art in CAD/CAM/CAE customization today.

ACED: EDITOR EVENTS

The AcEd library also contains objects that provide the applications developer with the ability to latch onto the drawing editor activities or events. The classes in the AcEd library contain function definitions that can be called as specific events occur at the editor level. Editor level events include starting/stopping commands and AutoLISP functions, returning from errors, and performing drawing file writes and reads. It is up to the applications program what actions might take place as the result of any given event. In most cases, editor events result in the applications program loading or storing information in the drawing or on disk.

These functions fall into a category known as reactors, or call back functions. In other words, they are run in reaction to some event. When the event occurs, the function is called to inform your application that the activity is taking place. Your program does not have the opportunity to cancel the activity; it's simply notified that it's happening.

The notification routines are typically available for when the editor activity is starting and when it is completing. There are also specific notifications for errors, and when an activity was canceled, for whatever reason. Your program can be alerted to the fact that AutoLISP is starting to evaluate something, that it has finished evaluating, or that it has encountered an error. It's your choice as to which events you are interested in knowing about. The following activities are just a few that are monitored at the editor level.

AutoLISP evaluator

AutoCAD command

DXF read and write

Drawing open and save

WBLOCK command

INSERT command

XREF operations

System variable changes

The need to monitor these events is application driven. The command start and stop can be used to track operator activity at a workstation. An automatic save utility can be created based on the number or types of commands entered. A drawing activity monitor can determine when operators are using particular blocks. The usage of events and what your program does with them will vary greatly from one application to another.

Event reactions are a powerful tool if used correctly. They can also be major headaches when used incorrectly.

Setting up reactors is discussed in a later chapter, but suffice it to say that the concept is similar to adding new commands. The memory address of the reactor function is supplied as part of the reactor setup as an ObjectARX function is called to let AutoCAD know about the reactor.

ACDB: AUTOCAD DATABASE OBJECT LIBRARY

Most applications edit the current drawing, so access to the drawing objects must be established. The drawing database objects in AutoCAD are accessed through the AcDb (AutoCAD database) library. The library contains object definitions for all of the AutoCAD entities and tables that make up a drawing.

A drawing is an object that contains multiple objects as seen in Figure 4.4. The objects inside the drawing object are tables, definitions, and lists of entities that make up the drawing. A program that accesses the drawing database must access the drawing object first, then work its way down to the target information.

ObjectARX has always allowed multiple drawings to be opened at any one time by creating instances of the drawing object for each drawing. In AutoCAD Release 2000

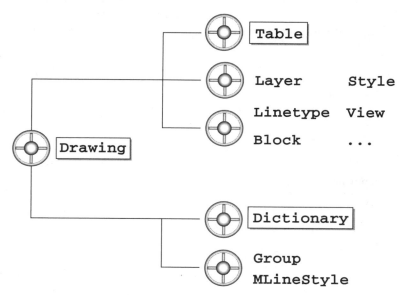

Figure 4.4 *Drawing Object Contents*

there can be more than one document open at a time. By going to the application object and then getting the document object list you can peruse the currently open drawings. Only one drawing is designated as the "current" drawing. Only the current drawing can be selected from in the AutoCAD edit session. In a multiple document environment, the operator must make the other document the current document before selecting something from it. In a single document environment (like AutoCAD Release 14), the current document is also the only document visible. One problem often encountered when working with applications migrating from AutoCAD Release 13/Release 14 to Release 2000 is that the current drawing can change while an operator selection is taking place. It is advisable to check if the selection made was in the same drawing you think it should be for your program to work. It is normally the malicious or curiously ignorant users who will push software not meant for such input and the solution is to simply check before jumping into the object or point right away.

Other drawings that are not viewed on the display can be opened inside a custom application for whatever purposes are desired. They cannot be viewed unless incorporated into the current drawing. An example of opening multiple drawings would be found in the AutoCAD INSERT and XREF commands where another DWG file is opened, and read into the current drawing.

To work with a drawing other than the current one, there is a function called "readDwgFile" which can be used to open a drawing as an object. Once opened, your program manipulates the objects within the second drawing using the same techniques as it uses on the current drawing; one of the nicer features of object-oriented programming inside AutoCAD.

Most of the time, an AutoCAD-based application works with just the current drawing and manipulates the entities found inside it. Access to individual entities is obtained through pointers. Pointers can be obtained from operator input where objects are selected on screen. Pointers are also found in various table objects such as the block table.

TABLE OBJECTS

Within the drawing object are table objects. Autodesk calls these "container objects" since they contain sets of data. The tables found in a drawing are for the layers, line types, text styles, dimension styles, view ports, and related AutoCAD details that are common to multiple entities in a drawing.

There are many-to-one relationships between the objects in the drawing, such as lines and arcs, and the table items. This is why there is a database level containing just tables. Another way to look at this situation is that there may be many entity objects that are on a given layer, but each object is only on a single layer at any given time.

Each table record is an object itself. There are specific methods, such as getName, for accessing information in the table record to obtain the object's name. Tables can be read sequentially in a program. An "iterator" object can manipulate each table object. The iterator object contains methods that iterate through the table, and retrieve records from the table in sequence. When writing reports that list the contents of a drawing, or when preparing to duplicate the current drawing, iterator methods can be employed. The two methods used most often are called "step" and "done." Quite literally, you step through the table until done. The method named "getRecord" is used to obtain the table record object.

Tables can also be accessed by the name of the entry you are interested in. A specific layer can be retrieved from the layer table by name and then manipulated. Each table object contains a method, which allows for searching the table for a specific name. There are two versions—one that searches to verify the existence of the name, and another that retrieves the table record object when found.

Creating new table entries is simply a matter of adding a new table record to the particular table object you want to expand. Procedurally, you first create a table record object for the table you want to append, such as the layer table. Next, you set the various record data items for the new table entry, such as the layer name, line-type, and color. Finally, use the add method for the table object to append the new table record.

THE BLOCK TABLE

The block table houses the names of the inserted blocks used in a drawing. Each record object in the block table contains an iteration object just like the tables. The iteration object in the block table record is used to walk through the entity objects that are associated with the block definition.

There are two automatic entries in the block table for each drawing. These entries exist when a drawing is first created, and will be found in every drawing encountered. They are the model space and paper space entity sets. Although they are not blocks themselves, they are treated at the same level as block definitions inside AutoCAD as seen in Figure 4.5.

To think of model and paper space as being blocks is somewhat confusing. From an operator's perspective they are certainly not blocks. And from the usage point of view they are not blocks. You would not want to insert the paper space objects into the current drawing.

From a drawing structure point of view, having model and paper space defined as block table entries make sense. They are simply collections of entity objects, just like blocks. The only difference is in the names and how they are displayed and manipulated within the AutoCAD system.

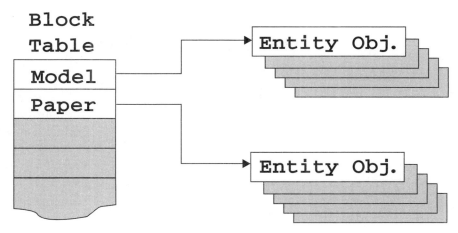

Figure 4.5 *Block Table Hierarchy*

To add a new entity object to a drawing, the first step is to select which drawing object we are addressing. Normally, that is the current drawing. From the drawing object the next step is to get the block table object. The model or paper space block object is accessed by name, and the new entities are appended. The basic relationship is shown in Figure 4.6. The drawing points to a table that in turn points to a record that then points to the entities. Seems like a lot of work to add a new object or access an existing one in the drawing, but it is not that much actual coding when using a language like C++. (There is an example of the coding required later in this book.)

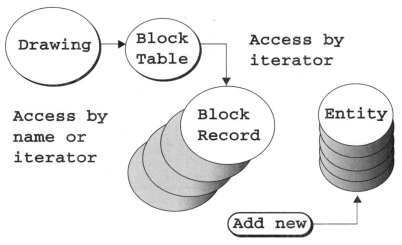

Figure 4.6 *Access Path*

ENTITY OBJECTS

Entity objects are the graphical information of AutoCAD. Entity objects are all defined in the AcDb library, which is the largest library provided with ObjectARX. AcDb contains class definitions for all AutoCAD objects that can be found in a drawing. This includes both the visible (entities) and the non-visible (tables).

Entity objects include lines, arcs, text, and other items that are used to construct the drawing. Working with entity objects is easy because there are methods available for updating all of the associated properties. Starting points, layers, and other properties are immediately available through these methods, which makes the resulting source code much easier to read and maintain.

For example, the methods getLayer and setLayer are defined for all visible entities. Once the entity is available to the program, one simply runs the appropriate method to manipulate the layer setting. If the variable named myEntity was associated with an open entity, the following C++ coding will set the layer to "0."

myEntity->setLayer("0");

In addition to methods for setting and retrieving properties of entity objects, there are methods for performing advanced manipulations. There are entity object methods defined for computing the results of transformations (scaling and rotation), obtaining the grip and stretch points, obtaining the object snap points, moving the grip and stretch points, redrawing the object on the screen, and computing the intersection of one object with another. These common entity functions provide a powerful set of tools for manipulating objects inside a drawing.

When learning about entity objects and how to manipulate them, the best advice is to look at as many examples as you can. They will provide the names of various methods that can be used, as well as show you how to manipulate the objects in ways you may not have imagined. Armed with some key names of methods, the next step is in the help and online documentation supplied with ObjectARX, or VBA, to learn more about the methods. When using the online documentation for learning, remember to select the "See Also" button for related subjects. You will be astounded at the number of useful methods that have been provided.

Entity objects are discussed in more detail in the next chapter. Entity objects are easy to manipulate using the methods provided. Virtually all of the functions that Autodesk uses internally for the development of new tools are available to the applications programmer. It's simply a matter of learning where to look for them.

OTHER LIBRARIES FOUND IN OBJECTARX

There are two more libraries found in ObjectARX. These libraries contain classes that are used by the other libraries for variable operations and manipulations.

ACGE: GEOMETRY DEFINITION OBJECTS

The other libraries use the geometry definition classes for the basic graphical elements required. These include points, vectors, and matrices. Being objects, the geometric library members contain methods for manipulating the data, as well as housing it. There are functions for performing basic matrix manipulations, such as dot product and cross product computations, along with a host of other useful routines. Most of the elementary mathematical manipulations that would be performed in a computer graphics environment can be found in the geometry library classes.

There are two basic object groups found in the geometry library. One set is for two-dimensional objects; the other for three-dimensional objects. For the most part, the same functionality exists between the two groups except that the three-dimensional group has more available.

Like the other libraries of ObjectARX, the geometry library is best learned through exposure to as many examples as possible. The ObjectARX samples and classroom exercises contain demonstrations of the most commonly used AcGe classes in action.

ACGI: GRAPHICS INTERFACE LIBRARY

AutoCAD objects are drawn on screen using functions found in the graphics interface library. This library contains classes that describe how to draw all the basic elements including lines, arcs, and text. It is important to note that objects are not added to the drawing database through calls to the graphics interface library. Instead, the database library (AcDb) uses classes from the AcGi library to render the objects described.

The AcGi library handles the processing of requests for things like slide viewing, and drawing temporary graphics such as the object snap point graphics seen in AutoCAD Release 14. AutoLISP and ADS programmers have used this feature before in the form of (GRDRAW), or ads_grdraw(), to create lines that aid the user. For example, an application could draw an arrow indicating where the user should look in the drawing in relationship to a problem. Or, a program might want to indicate potential input selections by putting an X mark on them.

Graphical elements created by the AcGi classes can be directed to all view ports at the same time, or to only one view port at a time. There are two sets of classes that can be used for that purpose.

When creating a custom object, the graphics interface library is used to solve the graphic generation issues surrounding the new object. Suppose the new object was a fastener such as a bolt. Numerous calls to the AcGi library will be required to render the bolt in either two dimensions or in three dimensions. More on this subject later when we investigate the creation of custom objects in ObjectARX.

SUMMARY

This chapter introduced ObjectARX in a bit more detail. The functions that make up the entire library are divided into various library areas based on the type of function. There are database functions, editor functions, graphical functions, system linkage functions, geometry functions, and other utilities related to working with AutoCAD.

A drawing is an object in which one finds more specific objects such as tables. Table objects contain various table member objects that may relate to other data objects in the drawing database.

Drawing objects and their sub-components are the most common objects manipulated by ObjectARX application programs. However, as seen in this chapter there are other areas where an ObjectARX application can interface with AutoCAD. From the editor library one can monitor and react to various AutoCAD command events and there is much more that we will be exploring in upcoming chapters.

CHAPTER 5

The AutoCAD Database—Entities

Entity access and manipulation is the goal of most AutoCAD customization tasks. In this chapter we will explore how the entity objects in the AutoCAD database are made available to ObjectARX applications.

ENTITY OBJECTS

Most applications written for use inside AutoCAD directly access the entity objects of the AutoCAD drawing database. Entity objects are the graphical primitives of AutoCAD such as lines, arcs, text, and circles. For instance, AutoLISP uses entity names and entity lists to provide such access. ADS uses result buffer chains which require sequential searches to arrive at the desired data elements. ObjectARX uses an object-oriented programming approach, which means the functions directly manipulate the individual entity objects in a uniform manner.

This chapter digs deeper into the classes, or object definitions used, to access the drawing database. It looks at the concepts of working with CAD database objects and how that concept is applied in the ObjectARX environment. An overview of the methods or functions provided for object manipulation is also presented along with a simple C++ coding example showing the concepts in practice.

WORKING WITH CAD OBJECTS

The object-oriented programming approach to CAD provides an elegant programming platform for the access of entity objects. The core of a typical customization program involves the manipulation of graphical components inside a drawing. The custom programs may need to generate new drawing components, change existing components, or take an existing drawing and prepare it for some other usage. Whatever the application, the core involves accessing the entity objects in the drawing database.

For quick review, entity objects are the Lines, Arcs, Text strings, and Circles that make up the drawing. A custom program using the methods or functions that are defined for the object classes may manipulate each of these objects. For each object type found in the AutoCAD drawing database there will be a set of functions that allow a program to directly access the parameters as well as change them. The functions will have the same name when performing like processes making it easy to learn the library and how to work with the objects.

Because these are objects, they know how to handle themselves. That is to say a Line object knows all about being a line. It contains properties that define the start and stop points, color usage, and other features that make up a Line object. The Line object also knows how to render itself on the screen, how to move or copy itself, how to compute intersections with itself and other objects in the database, and whatever else goes into being a Line.

Here's another way of looking at the programming situation involved. The object-oriented programming environment provides a class —object definition—which defines how to work with a particular object such as a Line. The class for a Line object contains definitions for various methods that access, change, or manipulate lines. When a custom program using ObjectARX accesses an object, the classification is first determined. If the object is a Line object then the custom program knows how to manipulate it using the methods associated with all Line objects.

ACCESSING AUTOCAD DATABASE OBJECTS

The ObjectARX libraries provide a comprehensive set of methods for working with AutoCAD drawing objects of all descriptions. In order to work with specific drawing objects, one must first determine what kind of object is to be manipulated. There are several ways in which an object's classification can be learned.

The simplest situation occurs when one creates the object in the first place, and therefore knows exactly what kind of object is involved. But, there are several alternatives to identifying the exact classification of an object if it is plucked from the database at random. The choice of which alternative to use is based on what one desires to do with the objects being interrogated, and how they are being gathered.

All sub-classes of the AcRxObject class have a descriptor object that can be used to identify the type of object being questioned. That means all entity objects in the drawing database have a way they can be identified as to what they represent.

In the AutoLISP and ADS environments, strings containing entity type names were tested to see if a match was made with a known set of primitive entity object types. This approach works fine in a static system where new objects are not created or derived from previous objects. In a dynamic environment such as ObjectARX, more is needed.

For instance, we may want to know if an object just encountered was derived from the Line object class. If so, we can use the methods for Line object manipulation to move the object around, even though our application may know nothing about the other features of that object. Before getting into object manipulations, we need to first understand how an object is identified.

OBJECT IDENTIFICATION

There are three ways entity objects can be uniquely identified to a custom program. ObjectARX programs use what is called an Object ID, so we will start our investigation at that point. The other two ways involve direct object pointers in memory, and entity handles that will be discussed during the exploration of the Object ID concept. All three identification methods can be used to reference a particular object in a drawing. It's a matter of what is available for use at any given time.

Each object in a drawing has a unique identification that takes on three different faces from the programmer's vantage point (see Figure 5.1). Even when the drawing is stored on disk and closed, each object has a unique code associated with it.

When stored on disk, an object's reference is called its "handle." A handle never changes for an object, and is never reused in a particular drawing. The Object ID is assigned when the drawing is opened. Object IDs will vary for a given object from one drawing edit session to the next. Direct object pointers for objects are assigned when the individual objects are opened and are only valid while the object is open. Once closed, the direct object pointer no longer points to the object.

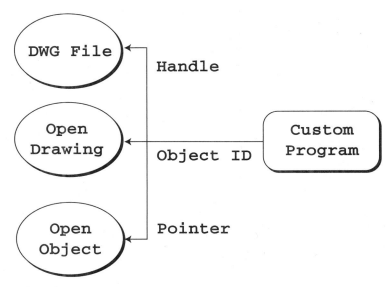

Figure 5.1 *Data Accessing*

Object IDs originate from multiple sources in a custom program. If the custom program creates a new object in the database, an Object ID is the result. When the contents of a selection set are retrieved, Object ID information is supplied. You can also walk through objects in a block definition, or other container object, and retrieve Object ID pointers to various items in the drawing database.

For those familiar with AutoLISP and ADS, the Object ID can be thought of as being the same as an entity name (type ads_name or 'Ename). It's the same as an entity name in that it is a unique identifier to an object in the current drawing. However, the mechanisms by which the two are represented in the computer are different.

As a consequence, it is necessary to convert ADS entity names into Object IDs for usage in ObjectARX. The converse is also true in that Object IDs must be converted to entity names in order to be used by the ADS library components. Since the operator uses the ADS library components primarily for selection set manipulations and entity selections, it's necessary to convert between the two styles of storage on occasion.

The following table lists the functions available for converting the various object identification types between each other.

Function Name	Does What?
acdbGetAdsName	Convert Object ID into an Entity Name
acdbGetObjectId	Convert Entity Name into an Object ID
getAcDbObjectId	Convert Handle into an Object ID
acdbOpenObject	Convert Object ID into Direct Object Pointer
getAcDbHandle	Get Handle given an Object ID

Table 5.1 *Table of Object Name Conversions*

An important concept to remember is that an Object ID, like an entity name, only exists while the drawing is open. The action of opening a drawing creates a new set of Object ID values to be used in the drawing. These Object ID values are created to house all the objects described in the drawing. As a result, a custom program cannot rely on the same Object ID being used for a given object when a drawing is opened at a later time.

HANDLES VERSUS OBJECT IDS VERSUS DIRECT OBJECT POINTERS

When the customization requires that an external system (outside AutoCAD) keep track of entities, the only solution available is to use handles. Only the entity handles remain the same from one drawing session to the next. As such they can be used to uniquely identify objects in a drawing. In other words, the entity handle is stored with the entity in the drawing file, and never changes. Once a handle has been used inside a drawing, it is never used again in the same drawing. Should an entity be deleted, the handle is retired at the same time.

On the other hand, Object IDs exist only when the drawing has been opened. Also, direct object pointers exist only after an object has been accessed. This situation exists to provide the best performance possible when accessing AutoCAD drawing objects. If your application uses a program outside of AutoCAD to reference objects, the handles can be used, as they will exist from one edit session to the next. When used, the handles are converted to Object IDs inside the program. Should the application only need to work with objects inside the current AutoCAD drawing, then Object IDs are used and the handles ignored.

Another issue is that drawing handles are maintained inside the system as strings of characters. This means that they may be used in applications outside of AutoCAD without any special conversion requirements. An example is a facility management database which links cost and maintenance information with a drawing. The database cannot store direct pointers into the AutoCAD memory space, nor can it store object ID data types. But the database can store strings. Thus, handles can be saved in the database environment. Use the conversion routines mentioned above to change

Access By	Data Type	Used When?
Handle	String	Lasts between drawing edit sessions and can be used in external programs such as databases and spreadsheets to reference drawing objects.
Object ID	AcDbObjectId	Created when drawing is opened and remains so long as drawing is open. Provides a naming system by which ObjectARX programs can access the entities directly.
Entity Name	ads_name	Same as Object ID but used in ADS modules for selection set manipulations and entity accessing.
Object Pointer	AcDbEntity	An object that is open for processing in an ObjectARX program is accessed using the object pointer. The pointer is used to direct the object handling functions to the appropriate entity object.

Table 5.2 *Object Access Options*

between handle and Object ID values inside an ObjectARX program in order to provide a bridge for entity linkage in a database.

So why doesn't AutoCAD just use handles all the time for entity access? The answer is speed. String comparing requires a substantial amount of computational time when compared to simply pointing at the object in the database. That is where the Object ID comes into play. An Object ID points into the drawing database to the record or information under scrutiny. While the drawing is open, the Object ID of any given object does not change. Object IDs simply deliver a faster way of accessing the information in a drawing.

Given an Object ID, one can construct a direct object pointer. Direct object pointers are used when an object is open for processing. In the programming world, they are used to reference the object methods for that particular object. In the source code the object pointers will be found before the method names. The following C++ statements result in the string variable object_layer being assigned the value of the layer in the object entObjPtr.

```
Char *object_layer;
object_layer = entObjPtr->layer();
```

There is another way of looking at the situation. The drawing currently open may be quite large and not entirely in local memory at any given time. The Object ID allows the custom program to locate the object no matter where it is, whether on disk or in local memory. When the custom program opens the object using the Object ID, a pointer is established directly to the object in memory. Behind the scenes the object may have been moved from disk into memory in order to accommodate the request, but the custom program does not know about the transition. The program does not need to be aware that the transaction between storage locations has taken place, all the program cares about is getting a valid link to the object for further manipulation.

From the AutoLISP/ADS perspective consider the Object ID to simply be a pointer just like an entity name and the direct object pointer more like the entity list or result buffer. But unlike AutoLISP and ADS, you will see the direct object pointer used for more than just manipulating the property values.

OPEN AN OBJECT FIRST

An Object ID is converted into a direct object pointer for use in a custom program by "opening" the object. Opening an object means that it is placed in memory where it can be accessed quickly (as opposed to out on the disk drive which would take longer) through the use of an entity object pointer. An object cannot be accessed using the database or entity methods unless it has been opened first.

To open an object in ObjectARX, the acdbOpenObject() function is used. Given an Object ID, the open function will create a direct object pointer. Custom programs must first open any object before accessing it.

The direct memory object pointer is of the class AcDbEntity, thus allowing the custom program to access the object using all of the methods known to the entity class objects. When used in the C++ programming environment, the variable assigned as the entity object pointer will appear before all of the function calls to the methods defined in the objects.

For example, the following program section takes a known Object ID and changes the color setting to a value of 3. The known Object ID is of the data type AcDbObjectId, and has been supplied from somewhere outside of this code snippet. Anything that starts with the characters "Ac" is from the AutoCAD library of names. The name "AcDbEntity" is used to declare a new variable of the type entity pointer. If you are new to C++, the double slashes are used to indicate a comment in the code.

```
// Declare the type of data - an entity object
AcDbEntity entObjPtr;
// Open the object in write or update mode
// knownObjectID is of type AcDbObjectId and was
    assigned elsewhere
acdbOpenObject(entObjPtr, knownObjectID,
    kForWrite);
// The direct memory object pointer is stored in
    entObjPtr when open completes.
// Now adjust the color index value for the object.
entObjPtr->setColorIndex(3);
// Close the object when finished with it.
entObjPtr->close();
```

Objects can be opened in one of three ways, two of which are used the most frequently; open the object for reading only, or for writing. The third is to open the object for notification, and that is only used if the object has reactors the custom program needs to speak with. Most of the time object manipulation is performed using the methods provided for accessing and updating the properties of the object. Although rarer to use, notification is available for consideration when building custom objects.

When opening an object for write, the custom application program must be the only program opening the object. This way, two or more custom programs cannot open an object at the same time and update it simultaneously. Instead, whichever custom program was last to open and update the drawing object, will have its changes preserved. If the object is open for any reason by another process, an error results during the open process when your custom program attempts to access it. It's possible to upgrade the status of an object that has been opened for read by one other process to be open for write by your program. After the program is finished with the object, the open can be downgraded to read only again.

A good programming practice to follow when updating entity objects: it is best to open the object, update it, and then close it as quickly as possible. This way the custom program does not interfere with other processes that may wish to access the entity object at the same time. The only exception to this concept is when working with larger groups of entity objects or performing many updates. Under those circumstances, the ObjectARX transaction manager should be considered. The ObjectARX transaction manager allows AutoCAD to delay updating the drawing database until all changes have been applied and your program decides it's time to commit the updates. The transaction manager approach is used by the block and polyline definition routines as an example of something that is written to the database with a final commit operation.

Other processes can open an object that is opened for read at the same time. That is, multiple custom programs can open the same object simultaneously, so long as it is for read only. In fact, the ObjectARX manual states that up to 256 independent applications can have the object open at once. Objects are opened for read if you are interested in obtaining the drawing data and not changing any of it. The custom program cannot modify the object when opened in this mode, it can only be accessed.

The open for notify option is provided so that a custom program can send messages to other objects. Since AutoCAD provides most of the methods ever needed for working with their own data objects, this option is primarily used to open a channel to custom objects. We will talk about custom objects in a later chapter.

DETERMINING THE OBJECT TYPE

Given an open object, the next step is generally to determine just what kind of object it is. There are four utility functions defined in the main runtime extension library (AcRx) that can be used to help classify an object's type inside a program.

It's important to remember that one can test not only for an exact match of a type, but also if something was derived from a particular class. Suppose a new entity object called a bolt-hole was defined by some other process. Suppose further that the new object definition was derived from a Circle object. Testing the object directly will find that it is a "Bolt" object and not a circle object. But testing if the object is either a circle or derived from a circle will find that the object behaves like a circle object, and can be manipulated as such.

Function	Does What?
isA	Returns the class descriptor for an object
desc	Returns the class descriptor for a known object (for comparison purposes)
isKindOf	Tests object against a specific class or derived class
cast	Builds new object of a specific type based on existing object, if it can

Table 5.3 *ObjectARX Functions*

The determination of which one to use is based on what your custom application needs. If you are looking for an exact match, then use the isA function with the object in question, and compare it against the desc result for a specific object. For example, the following line tests an entity object (stored in entObjPtr) to see if it is a circle.

If (entObjPtr->isA() == AcDbCircle::desc()) {

// entObjPtr points to a circle object!

}

On the other hand if your application will accept any object that is either a circle or has been derived from the circle object, the following code will suffice.

If (entObjPtr->isKindOf(AcDbCircle::desc()) {

// entObjPtr points to an object that is either a
circle or derived from one

}

Note that in both cases the "desc" function was used to retrieve the known descriptor for a circle. In a way, this is a constant value for comparison purposes. It results in a description of the object that can be tested against other descriptor data types. Descriptor data types are special to AutoCAD. About the only time you will use them is when testing for specific entity types.

MANIPULATING THE OBJECT

Once an object is opened, it can be manipulated and its properties accessed through a large selection of functions. The functions available are defined in the class descriptions for the entity object being accessed.

For example, a Line object contains its own methods for doing special operations with the Line. The Line object itself is derived from the AcDbCurve object that contains a set of methods, or functions, that are common to all the objects in the curve class.

The curve class is derived from the AcDbEntity class, which includes even more functions that are common to all entities.

As a consequence of this hierarchy, the Line object has a large number of functions available. More than most applications need, in fact, which is substantially better than not having enough.

When learning ObjectARX, a great deal of time will be spent searching the help files and header "H" files for functions that perform a specific task, but the most common functions will be learned quickly as they are the same for all object types. The class AcDbEntity defines a large library of functions that are available for all entity objects. These functions cover activities involving colors, layers, line types, object snap points, grips, transformations, and intersections. In other words, functions which need to exist for all the entity objects in a drawing. (The definition of the AcDbEntity class can be found in the file DBMAIN.H in the ObjectARX library.)

More object specific functions, such as setting the center point of a circle, are found in the class definitions for the entity objects. These functions are finer tuned to meet the specific needs, or to address the properties unique to the object. Consider the fact that a text object has different properties than a Line object. Each has data values that are different than the other, such as the starting and ending point of the Line object, as compared to the alignment points found in the Text object. (The exact definitions for the various entities can be found in the DBENTS.H and DBCURVE.H files provided in the ObjectARX library.)

Virtually any form of manipulation one could dream of is provided in the entity objects libraries of the ObjectARX tool-kit. That's because this is the same tool-kit as used by Autodesk to add new features to the system, and all that power is needed to deliver the quality product users expect. Since the entity manipulation functions are grouped in a fashion that is related to object-oriented thinking, the hardest part of learning the library is learning how to search for what you need. When looking for a function, try and consider whether it would be useful for all entity objects first, and if so, look in the AcDbEntity definition. If it is something that would have to be specific to the entity only, then look towards the entity class definition itself. After a little while, one becomes comfortable with the naming scheme used and rarely has to look hard for something new to use.

COMMON MANIPULATIONS

The most frequently used manipulations are found in the entity class. Since all entities in the drawing database are derived from this class, all entities share these routines in common. These include changing layers, colors, line types, and visibility properties of the objects. The following tables contain samples of some of the functions found in the libraries. They are presented in terms of the class they are used in, and some classes contain others. Please bear in mind that the following is only a partial list of what is available inside ObjectARX.

Function	Does What?
colorIndex	Entity color number
layer	Entity layer name
linetype	Entity line type name
linetypeScale	Line type scale associated with entity
visibility	Entity visible indicator
getEcs	Entity coordinate system
getGeomExtents	Geometric extents or limits of the object
getGripPoints	Entity points used when selected as a gripped object
getOsnapPoints	Entity points used when selected using an object snap mode
intersectWith	Find intersection(s) of object with another object

Table 5.4 *Access Entity Information—For All Entity Objects*

Function	Does What?
setLayer	Set layer for object
setColor	Set color code
setLinetype	Set the line type name
setLinetypeScale	Set object line type scale
setVisibility	Set object visibility
setPropertiesFrom	Duplicate properties for object from another object
transformBy	Apply transformation matrix to object points

Table 5.5 *Change Entity Information—For All Entity Objects*

All entity objects support the above functions. It does not matter if the object in question is text, a line, or a circle. They all share these common functions because they all share the properties for things such as layer names and color numbers.

Function	Does What?
getStartPoint	Starting point of object
getEndPoint	Ending point of object
isClosed	Determine if object is closed or not
getArea	Returns the computed area inside the object
getClosestPointTo	Returns point closest to another point on a curve

Table 5.6 *Access Entity Information—For All Members of Curve Class*

Within the entity class are a set of objects that are defined as being part of the curve class. The above table shows some of the functions that are common to all members. Members include entity objects such as polylines, lines, arcs, circles, ellipses, splines, leaders, rays, and Xlines.

Objects such as dimensions and text are not included in the curve group. Instead they have their own specific entity class definitions. Each class provides a set of functions that are used only for those types of objects.

Finally, each object has its own methods that are specific to setting the properties for the object. The Line object's most popular methods are listed in the next table.

Function	Does What?
startPoint	Set the starting point
endPoint	Set the ending point
thickness	Set the thickness of the object
getTransformedCopy	Create a new entity that results from applying a transformation to the existing entity

Table 5.7 *Entity Manipulations—For All Members of the Line Class*

Each object found in the AutoCAD database has a set of methods that permit direct manipulation and access to the properties related to the object. For example, the arc and circle objects each have a method for setting the center point and radius values. But only the arc has a method for setting the starting and ending angles.

CLOSE THE OBJECT WHEN DONE

When finished with an object, close it immediately. Closing an object makes it available for access by other processes—including your own. Even if the custom program only opened the object for read, close it as soon as the data extraction is completed. Keeping the objects free for access is important if the custom program is intended to run with other custom programs in the system.

Closing an object is accomplished by calling the close() method, which is available to all entity object derivatives. A common cause of memory leakage and erratic behavior in ObjectARX is failure to close the objects when finished manipulating them.

ADDING AN OBJECT TO THE DRAWING DATABASE

Adding new objects to the drawing is different in ObjectARX when compared with ADS and AutoLISP. In the latter two, either the command line is used by sending AutoCAD command sequences, or an entity list structure—result buffer in ADS—is created and sent to the entity create function. In ObjectARX the command line can be used as in the others, but a better approach is to simply define the new object and put it where needed.

To create a new object in ObjectARX requires that we first create a new instance of the object desired. Data is then supplied to fill in the object properties to match the application using methods that are normally available to objects opened for write mode. Technically, the object is now open for write and is under the same freedoms and restrictions as any other object that already exists in the database. In reality, the object is not in the drawing and as a consequence cannot be accessed by any other programs.

When the new object instance is created, a spot in memory is designated to hold the object, and defaults are filled in based on the current system setup. For example, the current layer name is inserted into the object when it is created. To change the layer setting, use the setLayer() method to establish the desired value. Specific entity properties that have no system default to draw from are generally supplied as parameters to the function that creates a new instance of the object type. In the case of an arc, the center point, radius, start angle, and end angle are all parameters used when defining a new arc object. Other properties like the line type, color, or layer assume the system default settings and to be changed will require a call to the appropriate "set" method.

Even though the object exists in memory and is considered "open," it does not exist in the drawing. To be included in the drawing, the entity object must be attached somewhere in the database. Normally it's added to the model or paper space definitions that are defined as blocks. The new object can also be added to a block—existing or new—in the same manner. A programming example showing how simply new objects are added to the database follows.

EXAMPLE PROGRAM SEQUENCE

To illustrate how ObjectARX all fits together, let's dissect a section of C++ code to see what is happening. In this example program we will add a line object to model space. The line will begin at point (1,2,0) and end at (2.5,4.25,1) on layer "DRW." The layer name "DRW" is assumed to already exist in the drawing and be the current layer.

The first step is to define the points. They are three-dimensional points so we will use the AutoCAD Geometry library definition. The points will be supplied in variables p1 and p2. In the C++ language they are arrays of three real numbers, but AutoCAD supplies a predefined data type called AcGePoint3d in the Geometry library as seen below. It is recommended to use the predefined data type as it improves the readability of the code and facilitates the locating of point variable definitions in source code.

```
AcGePoint3d p1,p2;
p1(0) = 1.0; p1(1) = 2.0; p1(2) = 0.0;
p2(0) = 2.5; p2(1) = 4.25; p2(2) = 1.0;
```

The above lines have established that the variables P1 and P2 are 3D points. They have also been assigned the values (1,2,0) and (2.5, 4.25, 1.0) respectively.

Given the data points, we can define the line object as being a new instance of the AutoCAD database LINE class. In other words, we set aside memory to store a line object in the drawing through the next sequence of code. The variable "pLine" is a pointer to the new line object.

```
AcDbLine *pLine = new AcDbLine(p1,p2);
```

The start and stop points are supplied to the new entity object when it is created. All other properties such as the layer and color of the object are based on the current system defaults. If you need to adjust the start and stop points the appropriate set methods can be utilized since the object is currently considered to be open for write.

Next, the block table needs to be accessed to obtain the pointer to model space. This is a two-step process in that first we need the pointer to the block table—the variable name will be "pBTable"—and then we need the pointer from the block table to the block table record with variable name "pBRec." In the C++ code, we first declare the variable types and then use the current drawing object to get the block table object.

```
AcDbBlockTable *pBTable;
// ObjectARX for AutoCAD R14
acdbCurDwg()->
    getBlockTable(pBTable,AcDb::kForRead);
//ObjectARX for AutoCAD R2000
acdbHostApplicationServices()->workingDatabase->
    getSymbolTable(pBTable,
 AcDb::kForRead);
```

Our next move is to get the model space record in the block table. This is where we want to add our new object. After getting the information, we can close the pointer to the block table object, as we no longer need to access it.

```
AcDbBlockTableRecord *pBRec;
PBTable->getAt(ACDB_MODEL_SPACE,
    pBRec,AcDb::kForWrite);
pBTable->close();
```

We are now ready to actually add the object to the model space. The variable pBRec contains a pointer to the block table record for the model space entities. Before appending the object, we need to declare an object ID pointer for our new object. This is so that we can close it after we are through.

```
AcDbObjectId LineID;
pBRec->appendAcDbEntity(LineID, pLine);
```

Last we need to close the entity and block table, as we are finished with them. The act of closing will flush the information in any buffers to the database, update the graphics on the screen, and clean up the memory it allocated for itself.

```
pBRec->close();
pLine->close();
```

As stated before, it's really not a lot of coding. The code just looks scary because the AutoCAD names are long to prevent overlapping with your own names. After a little practice, you get used to reading and writing them.

Another feature of the code example that makes it seem long is that the variables used were declared inline. If these variables were available on a global level in the code they would not need to be declared. That would result in four fewer lines of code required to add the new line object. Another way to reduce the code requirements for adding an object would be to already have pointers established to the block table items of interest or to use specialized subroutines for the adding of an object to model or paper space.

The number of lines of code required is a matter of programming style and standards that may be in place. Different programmers will undoubtedly have different techniques they may favor. The C++ language is very flexible which is why it is popular with many programmers.

SUMMARY

This chapter covered a lot of ground because the entity topic is the most frequently considered when writing ObjectARX applications. Starting from the notion of what an entity object is we then looked at how entity objects are addressed in the form of Object IDs. After comparing Object IDs to handles and object pointers, we then moved on to how to access existing object details. Opening the object and finding out what kind of object we may have been given was covered before moving on to manipulating the object details. Several tables showed some of the powerful functions available for manipulating AutoCAD database objects. We continued forward by looking at the aspects of adding a new object to a drawing presenting a simple program example in C++.

This has been our most technical chapter thus far. AutoCAD customization is a technical subject and requires some time to learn. There are many details and one cannot become an expert in the subject overnight. The important items to understand as we progress deeper into ObjectARX are the following. You should know what an Object ID is, and how it is used to get at the entity objects themselves. You should know that entity objects are first opened, then manipulated, then closed. And last you should know that the library of functions available for object manipulation can be referenced online with the ObjectARX online documentation.

CHAPTER 6

Reacting to Events

ObjectARX provides the ability for programs to react to events taking place inside AutoCAD. This feature provides a very powerful approach to integrating applications into AutoCAD. The steps involved in programming reactors are discussed in this chapter along with an overview of the notification system.

EVENT-DRIVEN PROGRAMMING

Most programs respond to the user. That is, the user starts the program and is asked to input data. Once the input is in hand, the program processes the data and creates output. The three steps of modular programming—input, process, and output apply to virtually all computer programs. However, nowhere is it written that the input must always come from the user.

Consider a computer system that monitors the water in a water treatment facility. The computer was originally programmed, then given parameters by an operator for data such as the acceptable water flow rate, and other measurements regarding water quality. That is the last time an individual is involved during normal operations. After being turned on, the input for the computer comes from the sensors that monitor the water flow, and quality checking systems. The computer is reacting to stimuli other than what might be considered normal for a computer. But when considered in the modular structure of input, process, and output, the process control program is following the norm. It's just that the input comes from another computer or sensor, which means it could also be coming fast and steady or sporadically as measurements are made. (See Figure 6.1)

Computer programs that react to the seemingly random input such as found in the process control system just described are sometimes called event-driven programs. An event-driven program does not have a singular thread of operation. Instead of the typical, clearly delineated steps of input, process, and output, event-driven programs contain modules that are run in response to some stimuli. Because of the modules

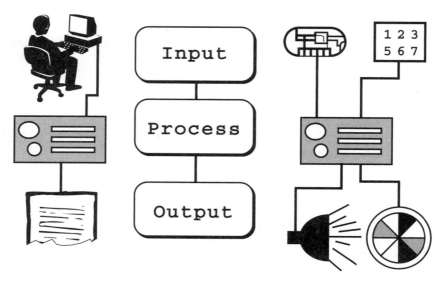

Figure 6.1 *Input Output Process*

involved, which react in an independent fashion, object-oriented programming strategies work well when developed in an event-driven environment.

Let's take a look at an example that virtually all AutoCAD users are familiar with: dialog boxes. The coding of a dialog box is an example of programming an event-driven environment.

From the user's perspective, dialog boxes do not have to be filled out in sequence, and as such, the program supporting the dialog box must respond to any of the input fields at any time. This is accomplished by writing a series of small, independent routines. They are then linked to the various dialog box components. Setting up the dialog box involves assigning functions—expressions in AutoLISP—and default values to the input fields, all of which is accomplished by calling subroutines from a standard library.

Once the dialog box has been set up, its interaction can take place and the program that created it calls a subroutine which handles user activities. Control is returned to the dialog box creation program after the user selects an item that forces it to exit. At this time, the main program begins to work with the input provided.

Linking is the crucial step that binds the custom program modules to pieces of the dialog box such as edit boxes and radio buttons. The program informs the dialog box of the location, or name, of the modules to run when a particular component is manipulated. Once the dialog box has been primed with all this information—as well as default data to display—computer control is passed from the program to the dialog box by calling a subroutine (named start_dialog in AutoLISP and ads_start_dialog in ADS).

Figure 6.2 shows the dialog box programming flow in either ADS or AutoLISP. The first step is to set up the dialog box by loading its definition into memory. Next, call back assignments are made. The call back functions are expressions, or functions, that are not evaluated, but assigned to be evaluated should a particular event occur with the dialog box; such as pushing a command button. Next, the dialog box is run and control is passed to the dialog manager until the user selects an exit option that returns control to the main routine.

It should be mentioned that, in general, very few applications would require much in the way of event-driven programming. If all a program must do is accept some input from the user, process it, and then generate some output, then event-driven programming beyond dialog boxes may not even come into consideration.

There are reasons why an application may need to react to events inside a CAD/CAM/CAE system. Events such as the user starting a command, or the addition of an object to the drawing database, can be of importance when developing custom CAD/CAM/CAE applications. For example, suppose a program was written that registers when drawing objects are changed at any time. One way to do this would be to start up a new program and have it compete with the other computer programs in

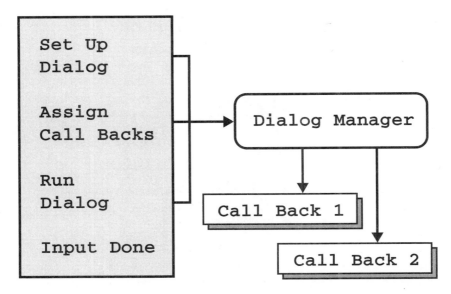

Figure 6.2 *Dialog Box Event Control*

memory for processor time, and when such time is granted the program would interrogate the drawing database for changes. Such an approach would be inefficient use of computer power and would result in slower operations. There is a better solution provided in the ObjectARX system. It is called notification.

NOTIFICATION

Another style of event-driven programming is when programs notify other programs as events occur. A host or main program carries out the notification after it has been told where to find the routines to notify. The functionality of notification must be provided in a main or host program, one cannot force it out of a program that has no provision for such inter-task communication. In ObjectARX, AutoCAD notifies our modules when events occur based on a set of standard events that are possible in the system.

Programming a notification-based system is essentially the same as programming dialog boxes, with one main difference. After the program sets up the reactor function links with the notification system—as well as the default values—the event set up program finishes. There is no call to a function such as the dialog run subroutine to invoke the event sequence handler. The events happen on their own, and the reactors in your library are notified by the notification system as the events occur.

Figures 6.2 and 6.3 help show how the two different event-programming strategies work. In Figure 6.2, a single program runs that performs a set up of the dialog box and

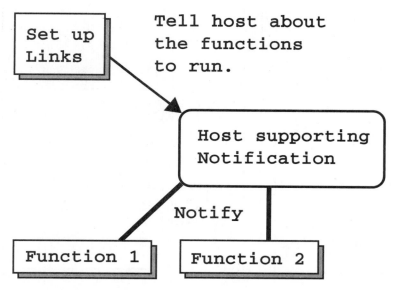

Figure 6.3 *Notification Event Control*

then proceeds to run the dialog. The dialog manager invokes call back functions 1 and 2 when the associated dialog items are changed. After the dialog box is completed, the program regains control and performs any activities desired now that input is finished.

In Figure 6.3 the set up program does its thing and is then finished. The main or host program then calls functions 1 and 2 when the associated events take place. Obviously, this approach allows for high degree of interaction between computer processes. As events take place, the notification host program can let the integrated programs know what is going on.

The functions stay in memory since they are part of a dynamic link library (DLL) program. If they were part of an executable then this approach would not work properly since they would disappear once the executable program finished. A DLL program waits in memory (or on the disk in a known location) to be activated, which is a far more effective way to provide a monitoring system. Another way of looking at the situation is that the DLL-based functions become part of the loading program and is simply available, as is the remainder of the host program.

OBJECT-BASED NOTIFICATION

By storing the information needed in objects, and using associated methods for data manipulation, independent program modules can communicate with each other. Suppose you create an object that keeps track of how many commands the operator has used in the system. This object would need to contain a method, or function, that initializes the object data elements. Initialization in this case would consist of setting a counter to zero and establishing a link between the notification system for command processing, and another method within the object. The second method in the counter object is the call back function that is responsible for accessing the current counter value and incrementing it whenever called. An additional method might then be used to access the current counter value and report it to the user or another program. The common ground is the object that allows the programmer to think of the data and programs together.

So how does the host program know to call the method associated with incrementing the counter? In addition, how does it know when to call it? The answer to these questions lies in making use of the concept of polymorphism in objects. Putting it in terms that are a bit easier to digest, an object might contain a call back function that uses a standard name that both programs know about in advance.

An illustration will help explain what we are getting at here. Let's say we have a notification object that is activated when a new entity has been added to the database. Let's further suppose that the notification object contains a public definition for a function named entity_added. In the DLL program we are writing that will react to the notification, the first step is to clone the class definition. Next we override the function definition for entity_added by defining our own version. Overriding is the act of defining the function name again in your own program thereby overriding the definition of the same name in another library. Most reactor functions that are defined to be overridden contain a null definition—immediate return—initially hence the only programming they will respond to will be whatever you write. The object pointer for our cloned class can then be handed over to the host program—AutoCAD. The host program knows about the function named entity_added and will call it whenever that event takes place. Should there be other potential methods that will react to events in the same object, they will do nothing because they have not been overridden with a new definition in the cloned class definition.

AUTOCAD REACTORS

Another term for a call back function is "reactor." Reactors are functions that react when notified by the host system. When provided by a vendor such as Autodesk, reactors are grouped together into class (object) definitions that contain methods for each of the potential reactors. When you clone one of the class libraries, you don't have to use all of the reactors found in the library. Instead you only want to override the ones that do something for your application.

ObjectARX provides a set of event notification options that can be used in a custom program to closely interact with AutoCAD operations. There are five basic event reactor categories in the ObjectARX system. An ObjectARX application can have any number of reactors associated with it, and can have multiple reactor functions associated with the same event. In other words, you can have multiple reactor functions that respond to an event such as adding a new object to a database.

The five reactor categories are for generic database changes, specific object changes, command-related activities, monitoring other ObjectARX programs in the system, and a general-purpose transaction system that is used when grouping operations together. Each of these categories is associated with different libraries in the ObjectARX system. For example, the entity-based reactors are obviously found in the entity library, the editor reactors in the editor service library.

The choice of which reactor to use is specific to the application in question. If an application needs to keep track of a specific set of entity objects, there is no reason to use the database reactors. Instead, one would want to specify a reactor that is directly linked to the objects to be tracked. Whenever the objects are manipulated, AutoCAD will notify the custom function so that the appropriate action can take place.

TRANSIENT VERSUS PERSISTENT REACTORS

Of the five reactor categories, there are two basic types that exist inside the system: persistent and transient. The difference between the two has to do with how long they exist in the system and who is responsible for maintaining them.

A persistent reactor is one that has been assigned to a particular object, specifically a database object, and, as the name implies, will remain attached to that object. A persistent reactor remains attached even after the drawing is saved and reloaded in a different edit session. As such, persistent reactors will only be found associated with database objects. This sort of reactor is typically used in conjunction with custom objects and is copied when the object is copied. Once a persistent reactor has been attached to an object, ObjectARX takes care of managing the memory for it, and other concerns such as what to do if your ObjectARX module is not available.

Transient reactors, on the other hand, are the most commonly used forms of reactors. Transient reactors can be attached to database objects but are not copied or saved with the drawing database. Transient reactors only exist while the drawing and ObjectARX program are both active. The reason transient reactors are used more frequently than persistent reactors are because they can be used in more circumstances and are available only when the custom ObjectARX modules are available. Persistent reactors are primarily used in conjunction with ownership hierarchy definitions. Whereas a persistent reactor can only be attached to an entity object, a transient reactor can be used to monitor the actions of the entire drawing, or system, as well as being attached to a singular entity.

One important difference to keep in mind when considering transient versus persistent reactors in conjunction with database objects is that transient reactors are not copied when the object is copied. This means that if you have attached a reactor to an object it can be copied without the linkage. Should you need to have a reactor for the copied object, attach a new instance of the reactor class to the new object when the copy operation is finished. There are reactor options that will inform our application when an object is being copied, and which would allow such a process to take place.

HOW TO SET UP A REACTOR

Programming a reactor in ObjectARX is not difficult. When getting started, the most difficult part is learning what reactors are available and what they are named. At the end of this chapter are tables listing the more popular reactors and what they do. These tables can be used to access more information inside the ObjectARX online help. The actual coding involved is not as much as one might imagine thanks to the nature of C++ and the Autodesk implementation of ObjectARX.

Reactors are set up in three basic steps. The first is to duplicate the classes that contain reactors. The second step is to define functions to serve as the reactor. The third is performed when the ObjectARX program is initializing and that is to inform the notification function about the reactor. In C++ this involves the creation of your own class based on an existing reactor class structure, as found in the header files for ObjectARX. Using the C++ keyword "class," the public functions that are involved in the existing reactor class are cloned and the names of the functions to be overridden are then declared. All that is needed in the declaration section are the names and the data types of the functions and parameters.

Once the class is defined in the source code—or an included header file—the functions can be defined in detail. The names of the functions must match those in the class definition for the specific reactors. These are the functions that will be called as the events occur and are the reactor functions themselves. All the function names are associated back to the reactor class that will allow us to reference the entire collection of reactors by a single name, or pointer, in the next part of the programming sequence. When the function is defined in your program, you are overriding the definition of that particular function.

The reactors are ready for use. The only thing needed is the code that links the reactors to AutoCAD itself. The time to inform the notifiers of AutoCAD about the reactor functions is when the ObjectARX routine is initializing. This is accomplished by creating a new instance of the class and saving the pointer to this new instance in a variable. The pointer is then passed to AutoCAD using the AddReactor() method in the current drawing object. The reactor set is now ready to respond, and AutoCAD knows about it.

One nice feature about using objects for reactors is that if we want to add a new reactor later, all that is needed is to define the function as part of the class, then code it and rebuild the application. No other coding is required. Thus, one can build a complete reactor-based application one reactor method at a time while learning how to control ObjectARX.

REACTOR OPTIONS
When working with reactors, the first step is to decide which reactor features will be needed by the application. The available reactor methods differ based on the type of reactor selected for use. Essentially, the choices are to be informed when an event starts and/or when it stops. When applicable, there are also choices for error recovery when the event was terminated due to an error in the process. For example, one of the editor reactors available will be started anytime a command is initiated by AutoCAD. Another reactor that is related is run when a command finishes, and yet another is available for situations where errors could be the result.

The reactors available in the ObjectARX system can be found listed in the header files. Details about the various reactors may also be found by searching the ObjectARX help system using keywords such as "reactor," and the names of the reactor classes. Of course, this is working under the assumption that you know C++ or are in the process of learning the language. Keep in mind that reactors generally fit into one of three groups: starting a process, ending a process, or aborting a process.

The processes being tracked by the reactors vary depending on the type of reactor involved. For example, the editor reactor knows when AutoLISP is starting, ending, or being canceled. On the other end of the spectrum, the database reactors know when objects are being added, opened, removed, or changed in the database.

REACTOR PROGRAMMING RULES

Due to the nature of reactors, and when they run inside the system, there are certain rules that must be followed when programming them. Reactors run as a response to various stimuli such as a command being activated, or an object being changed. Considering the fact that commands in AutoCAD can be initiated at the command line, from a menu, inside an AutoLISP program, inside an ADS-ObjectARX program, or through the ActiveX system, the rules of programming reactors that follow make sense.

- Reactors should not attempt to initiate new CAD commands, especially the editor reactors.

- Reactors should not attempt to get additional input from the user during normal command entry, as this is extremely disruptive to the computer and user.

- Editor Reactors should not use any ADS function library except ads_printf for displaying information.

- Reactors can report activity, but should keep it to log files and command line output only through ads_printf.

- Reactor sequences will change and be enhanced in future AutoCAD releases.

- Reactors cannot predict what reactor will come next.

- Reactors should respond only to the current event and not attempt to respond to other events.

- Reactors should not change the sequence of standard command operations by introducing new prompts or options.

- Reactors cannot veto an event.

Bottom line on the use of reactors is that they are started inside the AutoCAD system for a variety of reasons, and as such must keep themselves focused on the specific activities they are meant to monitor. A well-behaved reactor is only concerned about the specific items it is responsible for maintaining. Trying to do too much in a reactor will only lead to chaos and could adversely effect the performance of the AutoCAD system.

NO VETO POWERS

Reactors monitor only what is happening. They cannot nullify or change the event in any way. If an object is going to be erased, your application is powerless to stop the action. Instead, should that object need to stay in the drawing, your program can duplicate it and append the duplication to the database. From the user's perspective, there is an entity object that simply will not go away.

The reason why reactors have no veto powers in regard to events is that such a thing could quite easily result in chaos. Suppose, for example, that one function said to erase and another said to keep that object. The system would "lock up" and not return control to the user because one routine's actions would simply trigger the other routine. Thus, reactors can only watch was it going on and take additional actions, which in turn could result in further notifications being sent out. They cannot stop an event once it is underway.

Some may view the lack of veto power as a restriction in the reactor system, but really it is just a matter of establishing a set of rules so that all programs can work together.

REACTOR CATEGORIES

The following pages describe the reactor categories in a little more detail. Each category has its own unique set of methods associated with it. In some cases, there are more methods in a given class than an application will require. In fact, the editor reactor class is full of reactor functions that are rarely used. You can chose to use any, or all, of the methods from one of the reactor classes in your application. When defining the object, or class, in the program, only include the names of the reactors you intend to use.

The tables below list some of the reactors that can be found in ObjectARX. The names of the methods are provided to facilitate a search into the help system, or Visual C++

Reactor Function Name	Does What?
objectAppended	A new object has been added to a database
objectErased	An object has been erased from the database
objectModified	An existing object is changed inside the database
objectOpenedForModify	An object has been opened by a process and the open option was for update or write
objectReappended	A REDO command has resulted in an object that was removed via an UNDO being appended once again to the database.
objectUnappended	An UNDO command has resulted in an object being removed from the database

Table 6.1 *Database Reactor—Database Changes*

object browser, to find more details such as the exact parameter sequence, and return values that can be expected.

This set of reactors is provided for monitoring the entity object database in a generic sense. Any and all transactions with the database will result in a notification being sent to the reactors.

Database reactors could be used to count the number of new entities added, count the number of database changes made, and other tasks that involve the addition, changing, and removal of entities in the drawing.

Reactor Function Name	Does What?
modifiedGraphics	The specific object has been modified in some way

Table 6.2 *Entity Reactor—Specific Database Object*

When an object has been opened for modification (for write) and one of the methods called results in a change to the properties for that object, this reactor will be notified when the object is closed. The properties changed may or may not require a graphics update on the screen, but the AutoCAD system is going to do one anyway.

Entity reactors are specific to individual entities in the drawing. They are not persistent, and as a result are not saved with the drawing. When reloading a drawing into the editor, reactors for specific entities will have to be reinstated with the notification system.

Reactor Function Name	Does What?
rxAppLoaded	An ARX program has been loaded into the system and its objects are now available
rxAppUnloaded	An ARX program has been unloaded from the system

Table 6.3 *Linker Reactors—ObjectARX Program Comings and Goings*

Linker reactors are used to inform applications that other ObjectARX programs are loading or unloading. This is so an ObjectARX program can manage other ObjectARX programs—such as Mechanical Desktop—that may be part of a total system.

When writing programs that interface with other ObjectARX programs, there are additional tools, such as the function acrxLoadedApplications(), which provide a list of ObjectARX modules that are already loaded in the system. ObjectARX routines can load other ObjectARX applications, so if the module you need to interface with is not available, it can be loaded into memory.

One item of note is that an ObjectARX program cannot be unloaded by another program unless the application is unlocked. By default, ObjectARX programs are loaded in a locked mode, thereby preventing other ObjectARX programs from canceling them. The routine acrxUnlockApplication() will unlock the current application, allowing outside programs to influence the availability of the module. The need for this sort of module control is rare, but it can be used to load temporary functions, run them, and then unload them to release whatever resources they consumed in their operation.

Reactor Function Name	Does What?
commandWillStart	An AutoCAD command is commencing, the name of the command is available as a parameter
commandEnded	An AutoCAD command has just completed
commandCancelled	An AutoCAD command has just terminated due to a cancel request by the operator
commandFailed	An AutoCAD command has just failed due to an error
unknownCommand	The command just attempted is not in the valid command list and is invalid

Table 6.4 *Editor Reactor—Commands*

Editor reactors are the largest set of reactors available inside ObjectARX. The editor is the program that controls the interaction with the user through the menu system, command line, and AutoLISP interface. The editor can start many varied activities inside the system.

The above editor reactor functions relate to the AutoCAD command processor. They will be called as the result of an AutoCAD command being processed, no matter what the source of the command. The reactor signifying the start of a command will include the name of the command being started as a string parameter. Applications can use the start of a new command to control their own objects in a special way, such as hiding them from selection when the Erase command is started.

If your application requires that you add new objects to, or edit an object currently inside the database when a particular command is run, then special care must be taken not to interfere with the command underway. Because the database objects may be open for updates and other system conditions, it is strongly recommended that any database updates be performed when the AutoCAD command is finished. Even then, other transactions underway could result in problems, which is why the best solution is to avoid database updates inside of editor reactors if at all possible.

It is very important to remember that the editor reactors are activated when command interaction is taking place; so don't get too carried away with activities, as this will slow down the operation of the drafting system. Also, it is imperative to remember not to make calls to the ADS input, entity, and selection set functions during the editor reactors.

Reactor Function Name	Does What?
lispWillStart	An AutoLISP expression is starting
lispEnded	An AutoLISP expression has just finished
lispCancelled	An AutoLISP expression has been canceled by the operator

Table 6.5 *Editor Reactors (continued)—AutoLISP Related*

When an AutoLISP expression is evaluated, the reactors just listed will be called if defined in your application. These reactors allow applications to manipulate objects in preparation, or after the completion of an expression. The only expressions captured by editor reactors are those entered at the command prompt, via menu entry or user entry. Once the AutoLISP expression is underway, command-based reactors, or other object reactors, may be launched as a result of the activities within the expression, but no other knowledge of AutoLISP activities will be available until the expression is completed.

AutoLISP command function (defined with the C: prefix) names will be provided in the proper AutoLISP syntax with the C: prefix and parentheses. For example, the function (defun C:TEST () (print "A Test!")) will send the name "(C:TEST)" to the reactor.

ObjectARX programs that define AutoLISP callable functions should make sure all the functions are defined (using ads_defun) prior to AutoLISP starting a function. The best time to define AutoLISP callable functions from within an ObjectARX application is when the drawing loaded message has been sent to the main entry point subroutine.

Reactor Function Name	Does What?
beginClose	Drawing is about to be closed
beginDxfin	Drawing is about to import a DXF file
abortDxfin	DXFIN command terminated without success
dxfinComplete	The DXFIN command terminated with success, new objects appended to the database
beginDxfOut	DXFOUT command is starting
abortDxfOut	DXFOUT command terminated without success
dxfOutComplete	DXFOUT command terminated with success, new file created
dwgFileOpened	Indicates that a drawing file (DWG) is opened and being read into the system
databaseToBeDestroyed	Drawing is being removed from memory
beginSave	Drawing save operation is starting
saveComplete	Save operation is finished

Table 6.6 *Editor Reactors (continued)—File Transactions*

When files are being created or read, the editor notifies all the reactors listed above. Generally these reactors are used by applications that need to save data at the same time the drawing is being saved. When that is the case, it's best to have the program wait until the reactor completes to do the saving activity. That would mean using the saveComplete() or dxfOutComplete() methods to save your own application specific data to file. Should your data need to be stored in the drawing, you can attach the storage entities you require during one of the begin methods and then remove them during the complete or abort method calls. How you use reactors is up to you.

The file name used for the save operation is provided as a parameter in both the begin and complete reactors. If your application wants to save a file that is named the same as the drawing, but with your own extension, you can use the parameter string to construct the name automatically.

There are additional editor reactors that can be found in the ObjectARX help system, or the Visual C++ object browser under the class (object) name, AcEditorReactor. The additional reactors will notify applications when system variables are being changed, when blocks are being manipulated, and when subcommands are running.

The important thing to keep in mind when working with the editor reactor functions is that they will not always be called in direct sequence as events take place in the system. That is to say that the begin reactor will not always be followed immediately by the complete or error reactor. Other reactors may run during that same time period. You can rely on the sequence of begin to complete or abort for a given process such as saving a DXF file, but what happens in between the two can and will vary.

Reactor Function Name	Does What?
transactionStarted	A new transaction has started
transactionEnded	A transaction has completed with success
transactionAborted	A transaction has completed without success

Table 6.7 *Transaction Reactors*

Transactions are an alternative way of programming when handling groups of operations where one does not wish to clutter up the undo file system with many little changes. Other programs define transactions. These reactors are merely called to let your application know what is going on in the system. When a program—yours, AutoCAD, or another ObjectARX custom program—uses a transaction, these reactors are notified that the transaction is underway. When a transaction is running, you may want to handle operations differently, especially when working with custom objects.

The biggest advantage of transactions is that until the outermost transaction finishes, none of the activities are actually performed on the drawing database. Therefore, complex sequences can be programmed, and the decision to abort made in each function. Even though a previous function had completed its operations, a subsequent function can call off the entire activity for whatever reason, and the drawing will remain unchanged. This results in a much smaller undo tracking file and fewer reactor call backs as objects are held back to be committed all at once.

THE BASIC STRUCTURE OF A REACTOR PROGRAM

The following is a general outline for a reactor program set using Visual C++.

Define the class name for the reactor. All this statement does is set up the variable name myReactorClass to be a class.

class myReactorClass;

Create a pointer to the class by the name of myReactor and set it to a null value.

myReactorClass *myReactor = null;

Define the class methods, or functions, that we will be using; borrow the public definition of the reactor you want to use. The choices include AcDbDatabaseReactor, AcDbObjectReactor, AcRxLinkerReactor, AcEditorReactor and AcTransactionReactor. These names are defined in the header files for the ObjectARX library.

In the following example code sequence, the database option is used and the object appended reactor called out. By defining the exact same name as is found in the header file, we are overriding the subroutine definition with our own definition. In reality what we have done is borrow the other entire reactor overhead so that our application can be notified when an object is appended.

```
class myReactorClass : public AcDbDatabaseReactor
{
public:
 virtual void objectAppended(
 const AcDbDatabase* dwg,
 const AcDbObject* dbObj);
};
```

The next step in the program is to actually define the function that serves as the reactor itself. It will have the name objectAppended, but will also include the name of the object it is associated with. This is how C++ keeps track of what object a program is talking about. The following code shows only the subroutine definition and not the internal details of what happened when the reactor was called. That's up to the specific application!

```
Void myReactorClass::objectAppended(
 const AcDbDatabase* db,
 const AcDbObject *pObj)
{
 // do what ever we want with the
 // object ID in pObj.
}
```

With the reactor defined, the program now needs to decide how it will insert the reactor into the list of reactors maintained by AutoCAD. There are essentially two approaches to consider. The best approach, in terms of user friendliness and control, is to define a command that sets up the reactor. The command would have to be added to the command stack, which is a little more work than the second approach. In the second approach, the reactor is added to the reactor list when the entry point subroutine is called, and the ObjectARX application is initialized for the first time. The latter approach works fine for testing, and when you have an application where you do not want the user controlling how the reactors are inserted and used.

To continue our short example, the following code adds a reactor to the database reactor list.

```
if (myReactor == null) {
 myReactor = new myReactorClass();
}
// For AutoCAD Release 14
acdbCurDwg()->addReactor(myReactor);
//
// For AutoCAD 2000
acdbHostApplicationServices()->workingDatabase()->
    addReactor(myReactor);
```

From that point on, the reactor will be called whenever the event occurs. In this case, it will be called whenever an object is appended to the drawing.

To remove a reactor, the removeReactor() function is used in a similar manner as seen when adding a reactor. The add and remove reactor functions are methods which can be found in the drawing object. As an added note, the function acdbCurDwg() is a handy way to determine the current drawing object a program is working on. This function is not available in ObjectARX 2000 unless you make use of the following definition.

```
#define acdbCurDwg acdbHostApplicationServices()->
    workingDatabase
```

After the above definition is in place, the code looks the same in both ObjectARX versions for AutoCAD Release 14 and AutoCAD 2000.

CAUTION ABOUT REACTORS

Reactors are a powerful programming tool for application developers and should be used with caution. When you have reactors calling or resulting in the call of other reactors it is not hard to get the entire system twisted into an infinite loop. A word of advice when starting into reactors for the first time: expect the unexpected.

SUMMARY

The topic of reactors inside AutoCAD is new to most AutoLISP programmers and to many VBA programmers as well. ObjectARX provides a very powerful tool in that your application can know exactly what is going on inside AutoCAD at any time. It is important that any reactor programs written follow the rules so as not to cause any problems with other programs running in conjunction with AutoCAD.

In this chapter we learned about event-driven programming and how that concept leads to what are known as reactors. We discussed reactors in general terms as well as in programming terms for C++ developers. Reactors are established by supplying the addresses of the subroutines to run when the event is taking place. The different types of reactors available to the programmer were looked at with tables showing some of the types of functions available to aid in an application environment. Last, a C++ program segment was used to demonstrate how a reactor is programmed.

Writing programs that are based on reactors is not that difficult, just remember to play the rules of no user input or output, no commands, and that you can't stop something from happening gracefully. Reactor programming opens the doors to some interesting application ideas but it also opens the door for potential disaster. Do show caution when developing such programs and test each module created carefully.

CHAPTER 7

Making New Objects

The creation of new objects inside AutoCAD provides a vehicle to improve communications between different systems, address discipline specific tasks, and provide advanced solutions to custom CAD/CAM/CAE. This chapter introduces the steps involved in defining new objects and also looks into how custom objects can be utilized in different types of applications.

WHY CREATE A NEW OBJECT?

Why create an intelligent object when all one needs to do is add a layer name or insert extended data in the first place? What possible graphical object would I want to create that a block could not suffice for in the many applications of CAD/CAM/CAE I've been involved in over the years? These are questions I asked myself when first introduced to the concepts of custom objects. It was not until after careful consideration of what was meant by creating a new object, and what was specifically involved from a programming point of view, did I began to realize why I would want to create one in the first place. It all comes down to control and ownership of the objects involved, and keeping track of them in an application.

For example, suppose you wanted to create a new type of entity in the drawing that did not contain any graphic information, but contained only data. This new entity might mark the last time the drawing was changed, and by what user, as a way of keeping track of registered drawings in an electronic environment. The entity would be invisible to users, but would be written to the disk when the drawing was saved and manipulated by your custom ObjectARX application. When the drawing was opened in an AutoCAD session in which the custom ObjectARX application was not running, the object would remain invisible and unchangeable by the operator.

Another example of a custom object is a parametric symbol. Although blocks make fine parametric symbol warehousing systems, they are lacking when it comes to alternative viewpoints, grips, and stretching. A custom object that behaves in the man-

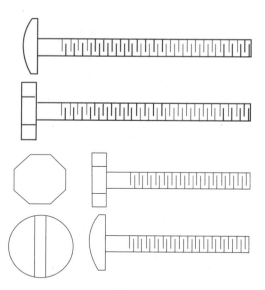

Figure 7.1 *Parametric Bolts*

ner needed by the application greatly enhances the usability of the parametric library. Consider a library of standard bolts as shown in Figure 7.1. The rendering of the bolt is different depending on the view—front or side or oblique—bolt type, and size. Additionally, when inserting the bolt one may want to select the head, the end, or some point along the shaft as the insert point. Finally, when stretching the length of the bolt, the thread diameter may not change until the bolt has reached specific lengths.

It's obvious that a symbol library built with these sorts of logical construction rules in them would be significantly better than a library of blocks. The symbols would be as intelligent as the coding that supports them. The symbols would know what a bolt is and how to draw it. Not only are the symbols more usable in the context of "being bolts" in a design, but reports such as bill of material counts, hole sizes, and locations, can be created by adding more functions to the custom object definitions.

Parametric libraries based on custom ObjectARX objects represent the next generation of part libraries. These libraries are smarter, and as a result can help the user in the usage of the libraries themselves. For example, given two plates of steel and a loading factor, it would be possible to have a program calculate the bolt required as well as the placement and selection of the nut that may go with it. In addition, these sorts of parametric libraries are well suited for tabular definitions of the parameters involved. This means that tools such as spreadsheets and databases may be used to maintain the lists of options.

CREATING NEW OBJECTS OVERVIEW

The creation of a new object in ObjectARX is not difficult from a conceptual point of view. The only real drawback to creating new objects is the amount of coding that can be involved for more extensive object definitions. The reason for all the coding is that when defining a new graphics object one must define a set number of methods that AutoCAD calls, as the object is manipulated in the system. These reactor functions can seem to be quite numerous and lengthy, especially for complex parametric objects. But putting the coding aspect aside for the moment, the creation of new objects using ObjectARX is quite easy.

In essence, new objects in ObjectARX are classes, which means they contain both data and methods, or functions. If your program simply defines a new class, or object, and uses it, then AutoCAD really does not know anything about it, unless you want it to know something about it. In that case, you must establish links between your custom object and AutoCAD. On the other hand, if you define a new class in your program as being derived from an existing AutoCAD class, then AutoCAD knows how to "talk" with it during runtime. It is up to you to define all the methods that AutoCAD will expect to find. This is where all the coding comes into play, defining all the custom functions for the new classes, or objects.

To get AutoCAD on speaking terms with a new object is a two-step process performed when the ObjectARX application is first loaded. Remember that the loading of ObjectARX application starts when the entry point function (named acrxEntryPoint()) gets the message AcRx::kInitAppMsg from AutoCAD. Typically, the initialization is handled in a function that is called by the entry point function when the proper message code is received. The practice of segregating the pieces of the entry point routine tends to make the code more readable and allows for numerous initialization steps to take place without crowding up the main entry routine.

The initialization routine is typically where one finds new commands being registered to AutoCAD, through a series of calls to the function acedRegCmds->addCommand(), as well as where global variables are established. Since this function is only called when the ObjectARX program is first being established into memory, this is the best place to perform that sort of activity.

In addition to registering new commands to the command table, the ObjectARX library function rxInit() is called to define new objects. This function initializes the new object and introduces it to AutoCAD by sending the name of the object to AutoCAD. Inside the rxInit() function is found a call to the ObjectARX library function newAcRxClass(), which is used to introduce the object description to AutoCAD. rxInit() is usually created through the use of a macro provided in the ObjectARX package. The macros can be found in the help system under the following names. Examples of their use can also be found.

Macro Name	Used to Create rxInit Call Sequence
ACRX_DXF_DEFINE_MEMBERS	For objects that are read and written to files; this one is used most often
ACRX_DEFINE_MEMBERS	For transient classes that can be duplicated
ACRX_NO_CONS_DEFINE_MEMBERS	For transient classes that cannot be duplicated

Table 7.1 *Macros to Define New AutoCAD Classes*

The important thing to keep in mind when using these macros is to watch your punctuation. String names do not require double quotes. More than one experienced programmer has stumbled on this syntax required by the system.

Once the new custom objects have been introduced through the calls to rxInit() for each object, the AutoCAD runtime class hierarchy needs to be rebuilt. The class hierarchy defines how the classes, or objects, in AutoCAD are related to each other. Most of the new objects being created are entities, however there are always exceptions. Calling the function acrxBuildClassHierarchy() accomplishes the object hierarchy rebuild operation. This is about the only time an ObjectARX programmer needs to be concerned with this aspect of the AutoCAD system. From this point on, custom objects are now available inside AutoCAD. Typically, the application will create new commands that will be registered to address the new objects in a specific manner.

Getting back to the coding problem, the number of methods required by a custom object depends on where it is derived from in the AutoCAD database. Typically, custom objects are derived from the AcDbObject, AcDbEntity, or AcDbCurve objects. These three objects pretty much cover anything you might want to create, and are explained in more detail throughout the remainder of this chapter.

When an object is derived from another object, it inherits all its properties. Putting this into programming terms, it means that the new object has all the data and functions that are found in the object from which it was formed. When a program redefines one of these methods, it is overriding the function definition. This is what is done with specific functions to support the custom object in the drawing.

OVERRIDING FUNCTIONS

The functions that are overridden when defining a custom class depend on which class the custom object was derived from. Overriding is the act of redefining the functions for our own use in our own class definition. When the program is compiled and linked, the new function definitions are linked to our objects.

The functions that are redefined have specific names. You may not select any name you want for the functions involved. In essence, these are reactor functions for AutoCAD to call whenever your custom object is manipulated in some way by the system.

The following table contains a partial list of functions, or methods, that can be overridden when defining new objects, or classes. Some are required, while others are rarely overridden due to the complexity of the operations involved, and the fact that the existing system does it just fine. This list is greatly abridged to provide a sampling of functions that are available to be overridden when defining a new custom object.

Function	From	Override	Does What?
dwgInFields	AcDbObject	Always	Reads data from file into object properties
dwgOutFields	AcDbObject	Always	Writes properties to drawing file
dxfInFields	AcDbObject	Always	Reads data from DXF file into object properties
dxfOutFields	AcDbObject	Always	Writes properties to DXF file
worldDraw	AcDbEntity	Always	Redraw object in world coordinate system using AcGi library functions
viewportDraw	AcDbEntity	Always	Redraw object in a viewport coordinate system using AcGi library functions
getGeomExtents	AcDbEntity	Always	Retrieve the geometry extents or limits of the object
transformBy	AcDbEntity	Always	Apply transformation matrix to object
getGripPoints	AcDbEntity	Always	Get points for grips on the object
moveGripPointsAt	AcDbEntity	Always	Perform change to entity graphics based on grip movements
getStartPoint	AcDbCurve	Always	Return the starting point of the curve object

Table 7.2 *Entity Functions to Override*

Function	From	Override	Does What?
getEndPoint	AcDbCurve	Always	Return the ending point of the curve object
isClosed	AcDbCurve	Always	Indicate if object is closed geometry or not
audit	AcDbObject	Most of the time	Checks new object when it is being added to the drawing database
intersectWith	AcDbEntity	Most of the time	Two function versions for computing the intersection of an object with this object
getOsnapPoints	AcDbEntity	Most of the time	Return a list of object snap points for the object
getStretchPoints	AcDbEntity	Most of the time	Return a list of points for stretch points for the object
extend	AcDbCurve	Most of the time	Extend the object to a new point
getArea	AcDbCurve	Most of the time	Compute the area inside the curve object
getClosestPointTo	AcDbCurve	Most of the time	Compute the closest point on the object to a given point
getDistAtPoint	AcDbCurve	Most of the time	Compute distance to point from object
getOffsetCurves	AcDbCurve	Most of the time	Compute an offset object from the current object
highlight	AcDbEntity	Sometimes	Highlight the object on the display
saveAs	AcDbEntity	Sometimes	Support for R12 and proxy object save structures
deepClone	AcDbObject	Rarely	Object is being cloned or copied
wblockClone	AcDbObject	Rarely	Object is being written to an external block
copied	AcDbObject	Rarely	Object has been copied

Table 7.2 *Continued*

Function	From	Override	Does What?
erased	AcDbObject	Rarely	Object has been erased
modified	AcDbObject	Rarely	Object has been modified

Table 7.2 *Continued*

For a detailed list of the objects and their methods, see the following header files located in the ObjectARX header file "include" directory. Examples for these functions can be found in the class labs and online help resources provided with the ObjectARX package.

DBCURVE.H	AcDbCurve methods and properties
DBMAIN.H	AcDbEntity methods and properties
DBMAIN.H	AcDbObject methods and properties

Table 7.3 *Header Files*

WHICH CLASS TO DERIVE FROM?

Because AutoCAD objects AcDbObject, AcDbEntity, and AcDbCurve are all related in a hierarchy, they share method names as well. That is, all the methods found in the AcDbObject can be found in both AcDbEntity and AcDbCurve. This is because AcDbCurve is derived from AcDbEntity, which is in turn derived from AcDbObject. The determination of which class you use to derive a new object from depends on what you need out of your custom object. The amount of work required in terms of programming is directly related to which class you select to derive from, as each has an increasing number of members associated with it.

Suppose your custom object is derived from AcDbCurve. It will have substantially more functions to override than if the object had been derived from AcDbObject. However, the number of methods to override should not dictate which class you use.

Which class to derive from depends on the nature of the object being created (See Figure 7.2). If the object is graphical in nature and must respond to the user's touch, then the AcDbEntity or AcDbCurve objects must be used. Objects derived from AcDbObject have no graphical component support. More often than not, AcDbEntity is sufficient to get the job done when dealing with objects that are to be drawn and

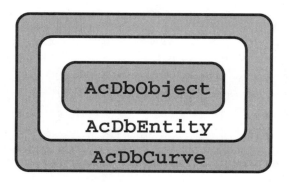

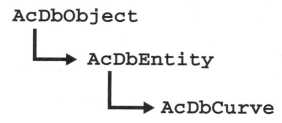

Figure 7.2 *Object Hierarchy*

manipulated. AcDbCurve is used when the object mimics a line, arc, or a polyline. AcDbObject can be used to create new objects that have no graphics. These objects contain only data and can be used to house information such as drawing logs and application parameters.

For the most part, parametric libraries tend to be derived from AcDbEntity since they will typically contain more graphics than objects derived from AcDbCurve. Objects well suited for AcDbCurve should have a start and an end point. Examples might include utility lines or fence lines. When deciding between AcDbCurve and AcDbEntity it's best to start with the entity level and then see if the object requires the added functionality found at the curve object level. For example, objects defined as derived from AcDbCurve will support an offset being generated, while AcDbEntity objects do not.

When creating new objects the best place to start is with the functions required for the highest level parent, and work your way down. More often than not, you will find that the entity level is as far as you need to go to implement a new graphic object. The curve class is used when you are dealing with a new object class that resembles a line, curve, or polyline. Many compound objects, such as the bolt mentioned earlier, are not well suited to definition as a member of the curve class.

WHY NOT DERIVE FROM EXISTING OBJECTS?

One item of important note is that the existing classes, or objects, for the graphic primitives should not be used for deriving new objects. Entity objects such as a line, arc, or text, should not be used to derive new classes. The reason is simple—they already work and there is little you can do to improve them except to add new methods.

Not much would be gained by deriving from the primitive objects since the worldDraw() and other functions would have to be rewritten anyway to accommodate your changes. Also, only ObjectARX 2000 fully supports these objects for the purpose of derivation. What this means is objects derived from many of them will not operate. It is best to use the more generalized parent classes and define custom objects properly. That is, with a proper description and support function set.

If you need to extend the abilities of existing entity objects by adding additional methods, there is a mechanism called "Protocol Extension" which is available for this purpose. AutoCAD uses an object that contains class descriptions to define all of the objects that are available in the system. Objects that are defined in the AutoCAD class description object list can be extended by having methods added to them. These methods are appended through the use of the addX() function and removed using delX(). Protocol extensions are something of your own design and must be called explicitly. AutoCAD will not call the functions that are part of the protocol extension when something is happening to the object. Database and entity reactors are used for this purpose.

SAVING AND LOADING DRAWING DATA

The most important functions your custom object must contain relate to the file storage and retrieval system. AutoCAD has no idea how to store or read your custom objects, so it is important that your program define a strategy for the properties to be saved. You need to program routines that define input and output from the binary drawing file—not as difficult as it sounds—as well as the ASCII text DXF files.

Actually, the file handling programming is quite straightforward and makes use of general-purpose functions. AutoCAD takes care of the tricky parts by storing an image of any custom graphics created on the screen. All your program has to do is store and read the object's data properties.

Data is written to, and read from, the binary drawing data file (DWG) using a simple set of subroutines. The functions dwgOutFields() and dwgInFields() handle the output and input of the parameters for the custom object. The good news is that the code is almost identical for all objects, with the exception that the object names are changed and the items actually written to the disk are different. Basically you write and read the data in the order you want. When outputting the object, the data is obtained directly from the properties and sent to the file using dwgOutFields().

When reading the data in for the object, the dwgInFields() function is used to obtain the data. Once read, the data can then be stored in the properties for the object.

When writing the data out, the first step is to prepare the object to be read and then call the parent class dwgOutFields() function. The parent class is the one you derived the object from such as AcDbEntity or AcDbCurve. After getting a successful return from the drawing out fields of the parent object, the writeItem() function is used to write each data item out. For the most part, you can use writeItem() to output the data in the proper type, no matter what you supply. However, on occasion you may need to force the data type to a long integer, in which case there is a special version of writeItem() called writeInt32() that may be used.

Reading data from the DWG file is just as simple as writing the data out. The first step of the dwgInFields() function for the custom object is to call the parent object dwgInFields() function. The function then proceeds to call the readItem() function for each of the properties. The read and write sequence must match up. If you add additional data to the output, you must update the input function as well.

DXF files are basically the same, however you must select the group code you want to assign to the object. Normally, the AutoCAD standard group codes are used in the numeric sequence they are typically defined in. That is, the numbers 38-59 are reals, 60–79 are integers and so forth. The main difference between DXF and DWG file handling is that DXF input should really support the possibility that the group codes are "out of order." Another program may read in the group code sequence and spit it out in a different sequence. It is your program's responsibility to be able to handle that situation. There are numerous examples available for processing DXF codes in random order that can be found in the Autodesk ObjectARX information kit.

DRAWING THE OBJECTS

A custom object defines what it looks like, or how it is rendered, in the worldDraw(), viewportDraw(), and saveAs() functions. This way an object can control what it looks like under different circumstances. Only the worldDraw() function is actually required for overriding. The other two simply use it if they have not been overridden in the custom object.

If the object is to appear differently when loaded in an AutoCAD workstation that does not have the custom ObjectARX modules loaded, then the saveAs() function can be used to define the exact graphics that will be used. The saveAs() function description of the graphical components of the object are also what can be used when the drawing is saved in the Release 12 format.

If the object is viewed differently in paper space than in model space, the viewportDraw() function can be used to describe the graphic rendering instead of worldDraw(). The distinction between which to use is controlled by the worldDraw() function. When viewportDraw() is to be used in a paper space viewport, the worldDraw() function must return a "False" result. If worldDraw() returns a "True" result then the viewportDraw() function will not be called. The actual calling of the viewportDraw() function is implemented by AutoCAD based on the returning value of worldDraw().

Function		What It Is Supposed To Do
WorldDraw	Required	Draw the object in the world view; return "True" if object drawn; "False" otherwise when viewports are active
viewportDraw	Optional	Draw the object in a viewport but only if worldDraw() returned a false result
saveAs	Optional	Draw the object for saving to a file; this rendition will be used when the drawing file is loaded in R12 or at a workstation that does not have the custom ObjectARX program running

Table 7.4 *Object Drawing Functions*

Inside these drawing functions you will add calls to the functions defined in the AcGi library. These functions create the actual visual representation of the object. Graphical primitives are defined in the AcGi library, which make it easy to describe custom objects made up of standard graphical elements such as lines, circles, and arcs.

The header file ACGI.H contains the definitions of functions used to draw objects in the graphic view ports. There are options for drawing circles, text, arcs, polylines, polygons, meshes, and lines. In addition to the geometry generators, there are also functions for defining graphic properties such as the layer and color of objects. The graphic properties are defined before the geometry is created, so that if you wanted to draw lines that were on layer "1" and red in color, you would first set the current layer to "1" and then set the color to red. The current layer and other entity traits are temporary and do not change the operator level drawing settings themselves. If you set the current entity trait to layer "1" in a module, it will not change the actual drawing's current layer setting.

The following table lists several of the functions that are found in the geometry library (ACGI) and what they do. You are strongly advised to review several examples showing these functions in action before coding your first one. Once again, this is only a partial list. There are several more functions that are defined in the ObjectARX help files, as well as the ACGI.H header file.

Function	What It Does
circle	Draw a circle
circularArc	Draw an arc
polyline	Draw a polyline, including arc and line segments with width
polygon	Draw a polygon
text	Draw text
ray	Draw a line
setColor	Set the current color for entity generation
setLayer	Set the current layer for entity generation
setLineType	Set the current linetype for entity generation

Table 7.5 *ACGI Library Functions*

The object drawing functions—worldDraw() and viewportDraw()—can be the most complex functions required in the definition of a custom object. Of course, this depends on how sophisticated the object is supposed to be in the first place. Simpler objects such as 2D details are much easier to automate than 3D rendering objects. Considering the bolt example again, a 2D representation of the bolt detail is nothing to program when compared to an accurate 3D model showing the thread details, and all the other elements of a bolt rendering.

When programming complex graphical objects, the best place to start is in the worldDraw() function. Create a simple version of the object that carries the associated data and shows it changing the basic outline shape of the object described. Sometimes this basic shape can be used in the viewport drawing function, so a good idea is to preserve a copy of the code at this stage in the development. Building the shape definition in stages allows one to test the effects of the parameter properties and get immediate feedback as to the usefulness of the graphical presentation.

Custom graphic objects are drawn based on the graphical primitives you use in conjunction with the parameters that make up your custom object. If the custom object is a bolt, you will need to store all the parameters of the bolt that are required to render it in the fashion you desire. Should those parameters include rotations or orientations, then the custom drawing functions must be programmed to support these properties as well. This portion of programming custom objects is where things can get out of control. The best advice that can be offered is to proceed slowly and one step at a time.

Sometimes the creation of custom objects involves a decision between the use of points, or parameters, to generate points. The decision is based on the kind of properties making up the parametric description of the object. The simplest custom objects to program and maintain are made up of points only. This option is therefore the favored one of the programming staff. However, there are times when a point-based structure may present problems.

For example, consider an object that is a cross made up of two lines. If the two lines are stored as four end points, then manipulation of the object is easy. The end points are simply changed using grips to form whatever type of crossing, or not crossing, lines the user wants. The programming required to make the cross object would not that be that arduous; the end points would be updated as needed and the object redrawn.

If the cross object contained properties such as the center point, rotation, and length of each edge, then things get a little more involved to obtain grip points and react to grip point changes. The points need to be calculated based on the parameters. The reason to consider this alternative representation in a serious way is that such an object definition holds to other rules of use better than one constructed of just points. In this case, the lines would always cross somewhere, which is more what one would expect of an object called a cross. The counter argument is to not allow the lines to not cross, but then we are adding more code. The decision to use points or parameters should be based on the rules of the object, and not whether it saves code or not. As stated earlier, custom objects can involve a lot of coding.

WHAT IF THE OBJECTARX PROGRAM IS NOT THERE?

After going through all the effort needed to create a new object, the question often arises as to what will happen to the object if the custom ObjectARX program is not loaded, or present on the computer? This circumstance will appear if you send a drawing created using your custom ObjectARX program to another AutoCAD user who does not have the custom ObjectARX program on their machine. When this happens, AutoCAD cannot rely on your custom ObjectARX program to handle these requests. Instead of ignoring the objects, AutoCAD turns all custom objects that do not have a runtime definition to support them, into what are called "Proxy objects."

A proxy object is an object that AutoCAD maintains. AutoCAD will attempt to display proxy objects, and will allow the user to manipulate them. The extent to which the user can manipulate the objects is controlled by settings made in the program when the object was first created and used. You control whether the user can change the proxy object, or select it at all, by setting proxy flags in your object definition. The PROXY_FLAGS variable is used to set up AutoCAD to be able to handle your custom object in the style that best suits your application.

You may wonder how AutoCAD knows to display your proxy object without your custom worldDraw() function being present? Actually, it's a classic case of some data slight of hand. The last time your object was defined in the worldDraw() or saveAs() functions, AutoCAD copied the contents sent to the AcGi—geometry drawing library—functions. Every time a line was drawn using the geometry generators, a copy was preserved in a graphic meta-file maintained by AutoCAD, and saved with the DWG file. If you want to disable that feature, and keep your drawing sizes smaller, the PROXYGRAPHICS system variable can be set in the AutoCAD session to a value that causes only a bounding box to saved. When the bounding box is used, your custom object graphics will not display on a system that does not have the ObjectARX module installed and running. Instead a rectangle showing the region covered by the object will be displayed in its place.

The original versions of ObjectARX referenced these proxy objects as "zombies"—a much more colorful and meaningful term.

Whenever your custom ObjectARX program is loaded, any proxies that exist in the current drawing are regenerated by calling worldDraw(), and are reinstated as supported objects in the system. When your program is unloaded, your custom objects once again become proxy objects. This is why I like the term zombies. When your program is present, the zombies live; but when your program is not there, the zombies merely haunt the drawing and the user.

HOW TO KEEP TRACK OF CUSTOM OBJECTS

There are a couple of ways to keep track of custom objects inside a drawing. If the object has been added to the entity class, including the curve class, then instances of it can be found using selection set filters. AutoCAD knows what the object is called and can perform the search based on that information, just like all other objects in the entity database.

Another way to keep track of custom objects is through a dictionary. A custom dictionary is an object that contains a collection of other objects. The objects inside a dictionary are referenced by names that you control. You define what the contained objects are, and what the names for the dictionary objects will be when creating and adding the objects to the dictionary.

This means that a dictionary can contain references to anything you want to store and access at a future time. When an object is added to a dictionary, a reactor is set up between that object and the dictionary. If something were to happen to the object—such as deleting it—the dictionary is automatically updated. All the dictionary actually stores, is a pointer to the object that was added to it.

So long as your object is in the drawing, and remains a member of the dictionary, the dictionary itself cannot be purged from the drawing. The member itself can be deleted from the drawing. If the dictionary is then empty, it may be purged.

Custom dictionaries also provide a method of storing non-graphical objects in a drawing. Because they are simply object references of our design, we only need to define the data which is saved in DWG—and depending on the application, DXF—files. The dictionary object storage system then provides a mechanism by which these objects can be addressed as a collection, and worked with as such. Normal object selection through the use of selection set filters will still locate these custom objects, but dictionaries are a better way of keeping them under control.

The creation and access to a dictionary involves some object manipulation in C++. It isn't too much, but some of it can be a bit daunting when first learning the system. The first step is to create a new dictionary instance of the AcDbDictionary object. Once it has been created it does not need to be created again. In most programs that access dictionaries, you will find a code snippet that first gets the list of dictionaries, then checks to see if it already exists. If not, it is created. Dictionaries can be named anything you want, but it is highly recommended that you keep them unique to your application. You would not want your application to collide with another application's dictionary definitions—which could be tragic. Here is an example code snippet that retrieves the list of named dictionaries and checks to see if our custom dictionary named "MYSTUFF" is defined.

```
AcDbDictionary *pDicts; //list of dictionaries
// get the list of objects, save in pDicts variable
acdbHostApplicationServices()->workingDatabase()->
 getNamedObjectsDictionary(pDicts),
    AcDbkForWrite);
AcDbDictionary *pDict; //dictionary object
if (pDicts->getAt("MYSTUFF", //get dictionary
   (AcDbObject*&) pDict, // put it here
    AcDb::kForWrite) == Acad::eKeyNotFound)
{ //it was not found, we must add it
 pDict = new AcDbDictionary; // new instance
 AcDbObjectId DictId; // dictionary ID
 pDicts->setAt("MYSTUFF", pDict, DictId);
 }
```

```
pDicts->close(); //close when done!
pDict->close();
```

New members are added to the dictionary using the setAt() method of the dictionary object you are accessing. As seen in the code above, the setAt() method from the dictionary list object was used to add the dictionary "MYSTUFF" to the list of available dictionaries. The next time we do a getAt() with the same object, looking for the same name, we will get the required linkage to the dictionary just defined.

Another way of thinking about this is that there is a dictionary that houses all the other dictionaries in the system. All it houses are the names. It does not know anything else about them; that is the application's job. When we want to access one of the dictionaries we get a pointer to it from the master dictionary. We then get pointers to the objects inside the dictionaries and do what we want with the saved information in the drawing.

Dictionary items are stored in the order they are appended. A sequential walk-through of the dictionary will result in the objects being returned in the same order they were added. You cannot adjust this order without destroying the dictionary, and rebuilding it anew.

To walk through a dictionary, one uses a function called an "Iterator." Iterators are classes that contain methods that allow us to step through the objects in a dictionary. The methods allow for testing to see if we have reached the end of the list, as well as one that sets the iteration to the next object. For example, the following code walks through the object dictionary defined above, "MYSTUFF."

```
AcDbDictionary *pDicts; //list of dictionaries
// get the lists of objects, save in pDicts variable
acdbCurDwg()->getNamedObjectsDictionary(pDicts),
    AcDbkForWrite);
AcDbDictionary *pDict; //the dictionary object
pDicts->getAt("MYSTUFF",
    (AcDbObject*&) pDict,
    AcDb::kForWrite);
pDicts->close();
// Now create a new iterator for our dictionary.
AcDbDictionaryIterator* pDictIter= pDict->
    newIterator();
MyStuffClasss *pMyStuff;
// So long as Iterator is not finished, do the loop.
```

```
for ( ; !pDictIter->done(); pDictIter->next()) {

    //retrieve the actual object from memory and put in
        //pMyStuff.
    pDictIter->getObject((AcDbbject*&)pMyStuff,
        AcDb::kForRead);


    //access is now available to the object MyStuff and
        //it's methods
} //end the for loop
// Remove iterator from memory and close dictionary
delete pDictIter;
pDict->close();
```

Iterators provide an easy method for navigating through dictionaries in a sequential manner. You can test the name of the object just retrieved to see if it matches one that you are looking for, or you can use the function getAt(), which is part of the AcDbDictionary object definition. This function will scan the dictionary and attempt to locate the specified name. If found, the getObject(), and other methods associated with iterators, can be performed starting at the position found. If a match is not found, a null result is returned from the getAt() function. Note that getAt() was also employed in the master dictionary list to locate the dictionary object in the first place.

Another method of moving about in a dictionary is to use the setPosition() method, which is part of the AcDbDictionaryIterator object. This function uses an object ID and attempts to position the dictionary iterator at the matching entry. The object ID being used is from the dictionary in the first place, so actually this is a tool for returning to a set position where you were before in an application.

Dictionaries provide a powerful tool for storing collections of our own designs but they may still be a bit daunting for those that do not wish to create their own objects just yet. Programming in this environment is a step-by-step process of learning and trying. However, once on the path, the journey is worthwhile and yields many surprises.

AN ALTERNATIVE FOR NON-GRAPHICAL CUSTOM OBJECTS—XRECORDS

Xrecords allow application-related information to be appended to a drawing without the associated hassle of creating new objects. Derived from the AcDbEntity class, the AcDbXrecord class is an object that is written to and read from DWG and DXF files.

The content of the Xrecord object consists of parameters only. There are no intrinsic callback functions associated with them. If your application needs to keep an eye on the activities of the Xrecord objects, you will need to attach reactors for them specifically.

Xrecords are easy to use when compared with the steps needed to create and maintain a custom object that contains data only. The steps are to first define a Xrecord object that will be added to a dictionary of Xrecords. Xrecord objects are not added directly with the drawing entities in model or paper space, but are instead attached to a dictionary object reference, or some other object in the drawing. In most cases, Xrecord objects are associated with dictionaries, which means that each object added must have a name. The name may be of your own choosing.

Using a named dictionary allows us to retrieve the Xrecord data by name from the dictionary. This can be very useful for storing global variables, or some other data from a program. AutoLISP can maintain Xrecords, and ADS as well as ObjectARX, making them quite attractive for housing critical data that must be saved with each drawing. Although the ability to access the objects and change the data from AutoLISP may not be a good thing for programmers wanting to keep their data secure, most applications need to expose their data to as many other applications as possible, and as such, Xrecords provide a valued service.

Xrecords have no size limitations like those found in the extended data system, which limits the application to 16 kilobytes of data per entity. You don't need to worry about how much space is being taken up by the data objects in your extended records. Although it can be hard to imagine needing that much space for extended information inside the AutoCAD system, it is possible.

When you create an AcDbXrecord object, you define a result buffer consisting of group codes and associated data in a result buffer list, much like AutoLISP and ADS. The function ads_buildlist() is useful in this regard. It creates a result buffer structure, which is what ADS uses for entity list representations. The reason is simple, the result buffer structure is supported in AutoLISP, ADS, and ObjectARX.

The data that is stored in the result buffer structure is whatever your application requires. This is a flexible solution as well, since new data elements can be appended to the result buffer as needed. At the same time, they may or may not be used in other areas of the application. The input and output routines for Xrecords do not have to be 100% in synch with each other, as in the case with a custom object. This may save additional programming time but could also present a problem in keeping everything at the same level.

There are two functions that put and get Xrecord structures. To save an Xrecord result buffer structure, the function setFromRbChain() is used. This function is associated with the AcDbXrecord class for the record you want to add to a dictionary. There are some sample programs that show this activity. They are located in the ObjectARX samples from the documentation.

To get an Xrecord object into a result buffer for access inside your program, the function rbChain() is used. Once the Xrecord is open through the use of the getAt()or getObject() functions, the rbChain() function may be run to move the Xrecord data into a result buffer of your choice. From there the data can be further retrieved using the resval function to get the type of data you need, based on the group code just encountered.

Even though Xrecords do not really fit into the category of custom objects, they do present an alternative for applications developers who are simply looking for ways to store associated data, but don't want to go through the exercise of building a custom object.

SUMMARY

The creation of new objects is the second most frequent area that many developers are interested in working. After learning about the existing AutoCAD entity objects many applications want to create new objects they can control exclusively. This chapter looked what is required to create a new entity object inside AutoCAD. Overriding the common functions defines how the new objects will behave inside AutoCAD. We also looked briefly at what will happen if your ObjectARX program is not loaded before moving into how to keep track of your custom objects inside a drawing using a dictionary. C++ code was presented that demonstrated the basic steps involved in defining dictionary elements. And to finish the chapter the Xrecord object was introduced since storing data specific to an application is a primary motivation behind building custom objects in the first place and Xrecords present a much simpler alternative that will work well for many applications.

CHAPTER 8

Object Interface Options for AutoCAD

Selecting the right tool for the job is not always obvious. AutoCAD supports a variety of customization strategies and picking which direction to travel can often be confusing. In this chapter we explore the various tools available and also lay out a road map for learning more about customizing AutoCAD.

COMMAND-DRIVEN VERSUS OBJECT-DRIVEN

Because operators, who enter commands via the keyboard, a menu, or some other input device, run AutoCAD, it's natural for one to want to customize it solely through the use of command sequences. A program so written is considered command-driven which means that it drives a series of commands to achieve what is desired. This tactic will work well for many activities that are needed to take place inside of AutoCAD applications and may be all that is needed to solve basic problems. (See Figure 8.1)

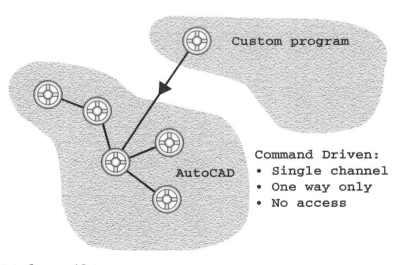

Custom program

Command Driven:
• Single channel
• One way only
• No access

AutoCAD

Figure 8.1 *Command-Driven*

There are problems with command-driven programs, the biggest being error recovery when something goes wrong in the command stream. For example, setting a constant text height in the current text style will cause AutoCAD to skip the text height request when the TEXT command is run. If the program assumes that the text height is not constant, and supplies a number, the parameter sequence is messed up and chaos could easily result. A program should check all possible settings that may effect the output of the command stream, but even that may not prevent all the possible errors that might occur.

Object level programming is very different from command-driven programming. In AutoCAD, however, they often get mixed together. When using ObjectARX for development purposes, there are times when one will be creating new commands, and care must be taken to provide options for others to use this command in a command-driven programming environment. This is where the two worlds often meet, so it's important to understand the motivations and requirements from the user side when programming. (See Figure 8.2)

Designers tend to look at CAD/CAM/CAE with different eyes, or at least they would like too. Often times, these CAD/CAM/CAE systems are designed for drafters who learn the vast command systems in order to carry their art into the electronic medium. As a result, designers find the same systems cumbersome to work with and are the first to look for better automation tools. When presented with a choice, most designers would prefer to work with objects that represent something more than simply lines and arcs.

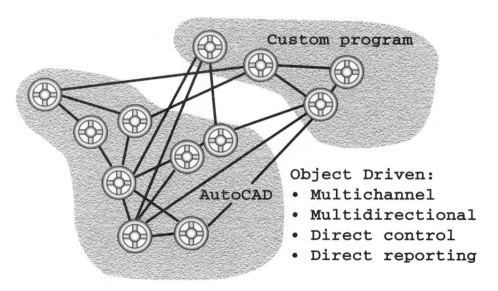

Figure 8.2 *Object-Driven*

EXPLOITING OBJECTS IN AUTOCAD

Before getting too deep into the object-level programming tools available, let's review what an object is, and how it can be used in AutoCAD. There are several kinds of objects and each has its uses in different areas of the system, or to achieve some goal in the customization effort. The entities themselves are objects that can be manipulated easier and in a more straightforward manner than earlier technologies. There are also objects that relate to system performance and what is going on at the time. Finally, there are objects of your own creation that are typically used to represent something from the designer perspective.

Entity-object manipulation is far superior to command-driven entity manipulation. About the only area where the command-driven alternative is preferred is creating new entities. Manipulation of objects, on the other hand, is better suited to the object style of programming. Not only are object properties immediately available, there are numerous functions that support the object for direct manipulation. When working in a command-driven environment, manipulation can fail if the system is not set up correctly. This is never a problem when manipulating AutoCAD objects directly.

When looking at past approaches to accessing AutoCAD entity objects, the newer object manipulation libraries are much easier to utilize. In an entity list—result buffer and DXF are roughly the same—manipulation of a LINE object, one must search out the group codes for the needed information. This format is strange to read for a novice, which could result in future support problems. Then, to replace the entity list information, one must substitute the appropriate group code in the correct place. Although AutoLISP makes this easy, it gets more difficult as multiple object components share the same group code; as in lightweight polylines.

A different area where objects are applied in AutoCAD customization has to do with the tracking and subsequent reactions to events in the system. Events such as database updates, command sequences, drawing saves, AutoLISP functions, grouped transactions, and other ObjectARX programs can be monitored though the use of objects supplied in the ObjectARX libraries. Normally one does not manipulate these objects, but instead clones them to attach the custom application to AutoCAD in a specific manner. When an event occurs, AutoCAD calls the routines associated with the cloned object, and as such they appear to react when something is going on at a critical juncture in the system.

Applications can also define custom objects that typically represent something specific to the application. These component objects are more than just blocks in AutoCAD, and can be very powerful tools for operators to use. Custom objects tend to be more application specific and, although they may represent a sizable investment in coding, they can return great rewards when implemented correctly.

Object-driven programming greatly exceeds the abilities of command-driven programming when developing any sort of application that has some depth to it. Depth means the application is fairly user-proof and offers multiple approaches to solving the same problem. Making a program user-proof is not that difficult in the ObjectARX environment, where every aspect of AutoCAD is under possible scrutiny by our applications. This way, an application can react on a real-time basis to the user and not have to wait to be reawakened to check out the situation.

PROGRAMMING SUPPORT FOR OBJECTS IN AUTOCAD

There are three primary programming language options that currently support AutoCAD objects. Depending on your background, it may be preferable to start with one instead of another. Each one has distinct advantages and disadvantages, and it's often confusing as to which is best for what sorts of applications. Most advanced developers use combinations of two or more to complete a specific application so it's a good idea to know about them all, even if you chose to only use one exclusively for development.

With the exception of one language, AutoLISP, the other choices are quite popular outside of the realm of AutoCAD. The external compiler options can be purchased through most good computer suppliers and there are several books on the subject of general programming for these languages as well. One problem with these books is that they rarely discuss AutoCAD, as that branch of programming is still considered highly specialized.

The AutoLISP language has several excellent reference and learning books available that address the subject at all levels. Having this large library when first starting into the programming realm can be quite handy, especially if one or more contain diskettes full of utilities.

THE C/C++ PROGRAMMING LANGUAGE OPTION

The C and C++ programming languages make use of the ADS and ObjectARX libraries to build applications. ADS is a library for use in C programs, but is becoming outdated as ObjectARX takes over. Most of the time, a C program can be converted to C++ without too much difficulty. During that transition, it can be converted into an ObjectARX program by changing the start-up sequences. The C++ programming language has a lot of depth and capabilities that may overwhelm someone learning to program, or picking up the language. The key to C++ programs is to understand how classes are used. Classes are how objects are defined, and once they are grasped, it's easy to learn how new things can happen in the language.

Advanced programmers who want to effectively use ObjectARX must learn and master the C++ programming language. However, they should also understand how

AutoCAD operators think and use the system. Creating new commands that are completely different than the basic AutoCAD system will most likely result in some user resistance. It's something of a paradox that experienced C++ programmers may have some trouble getting an ObjectARX application put together. Most of the time, some user guidance is required to make the system work effectively.

The primary tool of choice when developing ObjectARX programs is the Microsoft Visual C++ product. Visual C++ is a full screen, integrated environment that features a dialog box editor, object browser, and code editing system. The development environment is friendly but it might take some programmers experienced in other environments a bit of time and practice to get comfortable with it. ObjectARX online help files and documentation should be consulted as the best Visual C++ version to use, as well as other C++ development systems that are supported.

Microsoft Visual C++ contains wizards that help develop new programs. Since ObjectARX programs are actually DLL-type programs, that is the wizard option to select. Specific details on compiling ObjectARX programs can be found in the online help and documentation. For the most part, they involve a quick change to the main entry point name, and some simple parameter changes from the default settings. Then, you are ready to code.

To save time when developing applications for ObjectARX in C++, it's not uncommon to start with a template and fill in the gaps as needed. The samples provided with the ObjectARX developer kit are perfect for building a set of files that will serve this purpose. The best avenue of approach is to build a template for basic ObjectARX applications, another for programs which access the entity objects, and another for ones that define new objects, until you have the set you desire to build the kind of applications you need.

Although many programmers say that memory is cheap and we don't have to worry about how much space the programs consume, it's still a good idea to combine all the ObjectARX code for an application into a single program module. Unless the intent is to distribute the program in specific parcels, use of a single module will not only save space, but will also increase the throughput of the software. Even so, the parcels can be disabled through creative "secret codes" or some other activation signal for distribution as a single module.

THE VISUAL BASIC PROGRAMMING OPTION

The Visual BASIC programming language options are divided into two selections. There is Visual BASIC for Applications—-VBA for short—that is a sub-set of the Visual BASIC system, and runs inside AutoCAD. This option is in AutoCAD Release 14 for Windows. The other Visual BASIC (VB) option is to use the ActiveX library provided by Autodesk. VB runs outside of AutoCAD and uses ActiveX run time binding to link up with the AutoCAD application. There are a couple versions of VB available—regular and professional—as well as older versions that can be found at flea markets and discount computer stores. Watch what you buy! To program VB with AutoCAD Release 14, make sure you get a version that supports ActiveX automation programming. Earlier versions of AutoCAD supported some Visual BASIC integration through the DDE interface of Windows but this is not the same as the object-oriented approach available in the more modern version.

There are no significant differences between using VBA and VB except that the professional edition of VB includes support for more elements in the computer system. VB professional not only contains a decent database driver system but also comes with options for communications and additional controls, or dialog box components. VBA does not have as many controls as VB. However, it has the important ones you will need to get an application written. The main difference is that VBA is inside AutoCAD while VB is outside and driving AutoCAD through ActiveX automation. This means that some of the AutoCAD object references will differ, and the approach one takes to load and run a program is very different. But as far as interfaces, all of the Visual BASIC options provide an excellent tool for manipulating AutoCAD objects in an easy to program environment.

ActiveX automation is the newer, greatly improved version of OLE. ActiveX automation is a bridge between programs running outside of an application, such as AutoCAD and the application itself. The application, in this case AutoCAD, exposes a library of objects to the ActiveX automation system, itself an object structure, which then makes them available to any program wanting to talk to the application. The library of objects exposed contains both properties and methods that may be accessed by the calling program. These calls pass through the ActiveX automation system and into AutoCAD, where they are serviced. So one other aspect of programs written in VBA versus VB is that the VBA ones will be slightly faster.

VBA and VB allow for the same sort of object manipulation as found in ObjectARX. Objects are assigned to variables, which in turn open the door for function or method calls. For the most part, the function names are the same as found in ObjectARX. For example, all members of the curve objects contain the method startPoint().

Something that VBA and VB will not do when compared to ObjectARX is allow you to define a new object. Both VBA and VB rely on a set of DLL programs to run, and therefore are not an efficient way to define new objects. This is the one area where the languages fall short when compared to other object-oriented programming languages. VB has no ability to derive from another object and override its functions. Many would say that this disqualifies VB as a pure object-oriented language. While this may be true, the VB programming environment does provide for direct access to objects and the subsequent manipulation of them, which is more than enough for many applications.

Another aspect of VB and VBA that is different than ObjectARX concerns the complete support for reactors. Both VB and VBA provide access to the editor events for command start/stop, AutoLISP start/stop, and drawing load/save type operations. They do not, however, support any of the other reactors such as the database reactor system in the Release 14 implementation. For the most part programmers who need access to reactors are highly recommended to be working with the C++ ObjectARX development environment.

THE AUTOLISP PROGRAMMING OPTION

AutoLISP is the original programming solution for AutoCAD and it supports a variety of programming styles. AutoLISP can be programmed as a command-driven system, an object-oriented system, or a combination of both. Like VB and VBA, AutoLISP cannot be used to implement new objects, although it can access them. An improved version of AutoLISP that is available for developers from Autodesk is called Visual LISP.™ One special feature of Visual LISP is that one can convert the AutoLISP code into an ObjectARX program significantly reducing execution time. Visual LISP also contains more complete support for AutoCAD objects, including reactor objects. The basic approach to using the object-oriented routines in Visual LISP is the same as in ObjectARX. As a consequence, Visual LISP is an excellent learning tool for progressing into the ObjectARX environment.

When learning how to customize AutoCAD with no programming background, it's often recommended that you first learn AutoLISP, then progress into ObjectARX for more advanced requirements. Visual LISP serves as a bridge between the two environments giving programmers a taste of the extensive ObjectARX library with a slight boost in execution speed. For many this will suffice, as they have no need to interface with other processes or the operating system. When moving from AutoLISP to Visual LISP the speed of execution will improve about fourfold. The jump to ObjectARX will result in the execution speed being reduced several tens to hundreds of times.

AutoLISP is a unique language and has gained enough support that other CAD products have even begun to mimic it in their offerings. Most of these alternatives do not support objects in the same manner as the AutoCAD customization tools, and as a result are not recommended for purposes of learning how to program in AutoLISP. AutoLISP is unique since it is a programming language with an unusual syntax and has been greatly expanded by Autodesk to provide for operations inside a CAD environment. The newest generation of AutoLISP, called Visual LISP, does have the ability to latch onto and work with objects through its own ObjectARX bridge. The basics, such as reactors and the direct manipulation of entity graphics, are provided, which is more than enough for many sorts of applications.

A final note regarding the use of AutoLISP as a primary vehicle for AutoCAD customization is that it can be expanded through the use of ObjectARX. Not only can new objects be added, and then manipulated in AutoLISP, but ObjectARX can be used to define new subroutines for AutoLISP to call through the use of the ads_defun() function. This provides a seamless bridge between the two programming environments and illuminates a potential development strategy where a prototype can be built in AutoLISP. The prototype can then be migrated to ObjectARX in phases or the speed critical components converted to achieve the desired execution speed.

EXAMPLE APPLICATION IDEAS

The following are sample application ideas using objects. These applications are taken from real world AutoCAD situations and only represent some initial thoughts as to how object-oriented programming tactics may be applied. There are always other ways to implement such applications. It's hoped that the ideas presented here stimulate the reader into even more creative thought patterns in the world of CAD/CAM/CAE customization as well as object-oriented programming.

SURVEY NOTE REDUCTION

Most of the time survey measurements are presented in one of two ways. Either they are presented as output from an electronic survey instrument or they are notes written on pads of paper. In either case, the input information consists of straight-line measurements from a given point that may be referenced from another point. These measurements are called "shots" and are generally provided as distances, elevation changes, and bearing angles.

A note reduction program will accept these measurements and produce a terrain model of some sort. The model can be in the form of a map with lines showing the primary shots. Another desired output may be a contour map showing elevations and another may be an actual 3D terrain model suitable for rendering purposes.

An important item to keep in mind is that corrections to the terrain model will most likely be in the form of updates to the survey note values, or new shots provided that will override the older shots. Thus, one cannot expect the surveyor to update the terrain model in the CAD system once it is reduced to a series of contour lines on a map.

Custom objects can be used to represent various shots and reference points. Each object would carry identification matching that of the original survey note, along with any additional reference information about the shot series that may be desired. A command would be needed to import the survey notes into the object format. This command (or commands) would read the notes from the user or electronic data gathering device, and reformat it into the object properties.

A new set of operator level commands can then be created to read these survey note objects and create the map options desired. Of course, generating a contour map or surface from this sort of data is not all that simple to code. However, housing the survey notes as objects does present an elegant way to store the data in the drawing. As the survey notes are updated for whatever reason, the associated maps can be automatically updated.

Custom line types are another feature that can be taken advantage of when defining your objects. The generation rules for custom lines can be under your own control and will respond to scaling, and other placement rules more to the needs of the mapping user. Custom line types consisting of anchor point definitions are possible using the AcGi library in the worldDraw() function. You really can make it do whatever you need.

Finally, the use of custom objects in a mapping environment allows for many applications to be attached. These applications can be advanced models, such as drainage analysis, or simpler data object tracking systems for resources such as a core samples. In any case, the presence of custom objects can serve to greatly expand the intelligence of a map-type drawing, which could prove to be the foundation for other work in the civil engineering and architectural disciplines.

BILL OF MATERIALS SYSTEM

There are many approaches to take in building a Bill of Materials system. Extended data and block inserts are two options that are used most frequently by developers. In both cases, items in the drawing are tagged as reportable, and the Bill of Materials list is created after the drawing has been completed. When a change occurs, the Bill of Materials must often be regenerated from scratch.

In an object-oriented solution it would be possible to create a program that updates the Bill of Materials in real time. As items are changed, removed, or added, the Bill of Materials list would update as the events occurred. This would remove the need to rebuild the table anytime changes are made, and also help prevent errors as changes are reflected in real time.

There are a couple of ways objects can be employed to achieve this result. This application consists of two distinct parts, the Bill of Materials display and the items that are being counted.

For the Bill of Materials display you can use your own custom objects. That way the user cannot change them without modifying the model itself or by using your own custom commands that will police the changes made. A reactor for the situation where the custom object is removed from the drawing will be required so that the entire line associated with it can also be removed, and the entire report shuffled into new positions. The custom objects in the Bill of Materials display can be quickly found through the use of a dictionary search, or by a selection set. Once found, they can be updated to display the new information as the counts change in the design model.

Another alternative for the Bill of Materials display is to use blocks with attributes. The attributes can contain the count, and other values of importance to the display. An application program would then only need to know the Object ID values so that it can update them as needed. As in the custom objects, a selection set or dictionary can be used to locate the display items in the drawing quickly. Using a block with attributes may be more attractive to users as they can modify the appearance of the graphics themselves. As long as a standard attribute tag scheme is used, this approach should not present a problem.

Now there is the problem of objects that update the bill of materials in the first place. There are many more options available. Some may work better for certain situations than others. We will explore only a few.

The easiest to visualize, yet probably the one that would involve the most time to implement, is to use custom objects for the items to be counted. That is, if you are counting bolts and nuts, then use a custom object for the bolts, nuts, and all other items that are part of the assembly. When implemented correctly, this could be a beautiful thing and those that need this functionality should first look at the third party software market before starting into the project as a lot of coding will be involved.

When not using our own objects for reportable items, the question really becomes one of how to identify one item in the drawing from another. If a block library is involved, the situation becomes easier for automatic counting because each of these blocks can be tagged with Xdata or with an attribute to indicate they belong to a bill or materials count. A simpler approach might be to create a new block with attribute(s) that are the information destined for a Bill of Materials report. These generic, text only objects can be placed on their own layer and collected quickly to generate a Bill of Materials report. Both approaches just mentioned have been used by several developers in the past and are not too difficult to implement.

The custom program that updates the Bill of Materials display automatically can be implemented as a callback function, or reactor that is attached to the Bill of Materials objects. It would not matter what the objects are; a custom object of our design, or an attribute attached to a block. Whenever these values change, or are deleted, the reactor would run and update the associated Bill of Materials item. To figure out which item to update will require the use of a key, or tag naming system, that matches the Bill of Materials item with the display component.

The minimum amount of effort would involve using blocks with attributes for both the Bill of Materials display and as data points in the drawing for the items being counted. When the ObjectARX program is loaded, it would find the display items and build a list of the named objects involved and the Object ID values of the attribute objects themselves. Then, the data points can be located and the current values tallied up. If these values do not match the display, the user can be prompted that there is a difference and to select which one has the "correct value." The last step in the loading process would be to attach a set of database reactors to the current drawing object that are notified whenever new entity objects are added or removed. The automatic update program would then be ready to run. The reactors just added would look out for changes involving the data points, and when they occur it would update the display value of the attribute in the Bill of Materials.

ROADWAY DESIGN AND LAYOUT

Many steps are involved in the creation of a set of plans detailing the design of a new roadway. There are a few particular problems that face designers of roads when they are trying to put their thoughts in plan form using a CAD system. The coordinate system of a roadway is actually best thought of relative to a centerline that runs along the entire course of the roadway. The CAD system, on the other hand, makes use of a grid-based coordinate system. The first problem to solve is how to have the two systems work together.

The centerline of the highway from which other details are located in a design is often called the highway alignment. It originates from survey information. Each alignment point contains a station number and an elevation. The station number indicates how far down the highway that station point is located. If there are rapid transitions in the terrain, there will be many data points and if there is little change there will be fewer stations defined. Alignment points may also contain information about the next segment of the highway. For example, an alignment will typically indicate that a curve is starting, and to what direction and degree of curvature, or it will indicate that the next section is a straightaway.

To define the highway alignment in AutoCAD, one can use a polyline with extended data attached to the vertices, indicating the station numbers and parameters used to generate the next segment. This solution will run into problems if the user edits the polyline outside of the highway alignment information control system. In order to correct this situation, one would have to translate the polyline coordinates back into highway alignment parameters, which may or may not result in workable values.

Another way of implementing a centerline object would be to define a new object that is a highway alignment object. It would consist of a starting datum point and a series of alignment definitions. The worldDraw() routine associated with this new object would essentially draw a polyline connected at the data points. Additional stations could be inserted using custom commands that access and manipulate the parameters of the new object. Because you can use primitive objects from AcGi to render the highway alignment, proper centerline drafting strokes can be used, and the drawing can be created in 3D if desired. Internally the alignment object would store station numbers and alignment control specifics such as the type of segment ahead, degree of curvature, and the elevation at the point described. The points used in AutoCAD would be calculated based on these parameters, however, there is no need to store them. Like a circle object, you only need to store the parameters that are used to calculate the points.

Details for highway layouts are defined relative to a station number on the highway alignment, an offset value, and an optional elevation change. The elevation may or may not be part of the offset, and may instead be the result of applying a desired slope. The offset value is used to locate the detail point to the left or right of the alignment. This offset value is a specific value. The intersections from the previous, and to the next detail point, need to be calculated from these measured points. If the detail being described is a curb line, then it must follow the basic contour of the alignment, maintaining the same offset as described, or holding a constant rate of change between station detail points; no matter what geometry is described between them.

Thinking in terms of objects, the centerline definitions would be one type of object set which can be related through a dictionary to maintain the sequencing. The details are also objects that define the offsets, and can be associated together in a dictionary object so that our custom program can keep track of them easily. When an alignment is changed, the worldDraw() function would call all of the attached details to update them all at once, providing the highway designer with a tool they can use to quickly adjust highways in their own terms.

This system can be implemented without custom objects but not to the same degree as just described. Using reactors, changes to the alignment graphics can be used to trigger automatic changes to the offset graphics but the resulting edits would be slower than generating them as custom objects, and one would end up doing just about as much programming development work either way. The custom objects can be used to calculate other details that are useful to the highway designer such as the cross section drawings and determinations of earth removal or fill. They may also be related to analysis of rain runoff and catch basin requirements.

Unlike polylines that may have to be reduced back into the highway alignment parameters, the custom objects would carry these values all the time and allow for easy edits of the important data. In addition, the intelligence of the highway alignment program can grow as you are ready to expand them. For example, it would be possible for the alignment layout program to recognize when a route will require a bridge or an extensive amount of cut and fill. When these conditions are found, the program could then ask for further guidance from the designer and even accept alternative route definitions for purposes of experimentation.

A NETWORK

Network-type applications are found in a variety of disciplines—electrical networks, plumbing networks, HVAC networks, computer data networks, and information flow networks. Instead of tackling one in particular, let's take a look at the general problem of constructing a network and maintaining it. In particular, we want to make sure the network is connected properly and produce a report detailing all the connections.

There are several ways to solve this problem. Using blocks with attributes, or extended data along with the associated programming required, a good networking solution could be constructed that fulfills all of the base requirements. The application can be extended with ObjectARX through the use of reactors which would check the network connections as the users attempt to move components about. Let's look at the application from an object-oriented programming point of view and we'll see that a much more robust application can be constructed from such a model. We will start by defining the rules for the objects involved.

In this application there are two basic types of objects: the components and the connectors. The components are the things being connected by the connectors. Both the component and connector objects contain an object that is a connection. A connection can be either input or output. Components have input and output connections to which the connectors attach and may have any number of connection opportunities. Therefore, one of the other properties that will be found in the components is how many connections it has available. We will not go into the C++ class descriptions of these objects now, just remember that an object can contain other objects. In this case we are making use of that fact to keep track of the connections.

Next, focus on how the objects would appear on a conceptual level. To do this, define a set of rules that the objects live under. For example, a connector can only connect to two connections, one at each end. One connection must be designated as input and the other as output. Connectors can only connect to components and not with each other.

The properties of a connector would be the ID tags or names of the components that it connects with, and the length of the connection between the components. This length may or may not be reflected in the drawing of the connector. When creating a ladder diagram, or star diagram of the network, the connector lines drawn will bear little resemblance to the actual lengths involved.

The components are very much like the connectors except that they have no length value and can have any number of connections. The types of connections will depend on the component. The component could have input or output only such as a real-time sensor or alert lamp. Components will most likely have more abilities and information based on what the actual application may be. An electrical network may contain power draw information for each component, a computer network diagram may contain information about the peripherals available at each component, and an HVAC network may have flow rate data.

With these objects and rules one can create a library of custom objects that are derived from them for purposes of the graphical representation of the network. Another way of saying this is that we will define new components as being children of the component class. The component class family starts with a basic component type that contains a connection object. There are more advanced components with multiple connection objects inside, and there could even be components with components inside of them. For the drafting side of the work, each new component object would have an associated worldDraw() method that defines how it appears when used.

If this object structure for network design is employed, it will permit the network to be analyzed for accuracy, with missing connections highlighted quickly. Not only would the graphical components behave like they should, but they would be somewhat intelligent as to their purpose and usage.

One might wonder if such a system would take longer to develop than a system that was not based on objects? The answer is that the initial development of a non-object-based solution might be completed faster, however, it would be difficult to expand into a more automated system. A nice feature of object-oriented programming is that it allows for easier growth as new ideas and concepts are formed. This is important because virtually any programmer will agree that users come up with more ideas once they see the "initial version" of a program.

A MOTOR

It is surprising to some how object-orienting programming can be applied in virtually any discipline. To illustrate, look at the work involved in designing an electric motor. This example is greatly simplified and provided just to make a point about the use of objects in such an environment.

An electric motor —like Thomas Edison used to build—has a central shaft that spins a magnet. Coils of a conductive material such as copper surround it. The neat thing about motors like this is that they can be applied in two ways. In one, the shaft spins using some exterior force such as a waterfall or windmill. The result is the generation of electricity. On the opposite hand, we can supply electricity to the coil and spin the shaft to drive a conveyor, or some other device.

How would object-oriented programming concepts be applied to the design of electric motors? First, look at the way human designers go about the design of a new motor. A designer would know what power was available or desired at each side of the motor, and would know other parameters such as start and stop time requirements. Starting with the shaft, the size would be dictated by what is attached to the outside of the motor. From there, bearings to hold the shaft can be determined. The last object to consider is the magnet attachment and how large it should be. The size of this portion will depend on the electrical power requirements.

The objects would be the parts that specifically make up the motor. From a simplistic level, these would include the shaft, the bearings, the magnet, and the coil. The motor will have a shaft that is suspended between two bearings and a drive system. The size of the load controls the size of the motor, which dictates the size of the bearings needed; and that in turn defines the size of the shaft. If programmed properly, an entire motor design could be created from object information only. Once an automated design system has been achieved, it is not that difficult to conceive of an automated machining system picking up the object data and manufacturing the custom motor.

STRUCTURAL ANALYSIS

Another discipline where objects can be employed is in the study of structures. Structures can be simple, such as a wall; or more complex, such as a large building or bridge. In most cases, structures are used to hold things up, which means that they carry some sort of a load. The load is translated through the structure to the base of the structure, which needs to be strong enough to hold everything, including the structure itself. When figuring out how strong the members at the bottom must be, one needs to start at the top with a known load, and then calculate how that load is transferred through the remainder of the structure. Each member of the structure will transfer the load in a different way and will itself add more loads—in the form of weight—to the overall structure. How a member transfers a load is dependent on how the member is attached to the structure, but this is beyond the scope of our current mental exercise.

A structural analysis system can make use of the object-oriented programming concept. Start with each structural component that can be defined as an object that has load, or weight, and other related properties such as how it can be attached to other members. Then begin with the simplest of structural types and build a library of members that can be combined to form a more complex structure. Looking at bridge design as an example, there are standard structural components such as "K" transverse sections. These can be thought of as objects also. As each new object is defined to the system, one of the methods it would contain would calculate the translation of load through the member. If a "K" frame was made part of the library, the loading can be calculated through the entire frame and not require a total recalculation of the frame members. This is where an object-oriented system excels. Since objects can be used to define other objects, it's not that difficult to add new members to a library once the basic library structure has been defined.

LEARNING MORE—WHERE TO GO NEXT

So where do you go from here? A lot depends on your background and knowledge of AutoCAD, and programming in general. Even more depends on how much time and interest you have to invest in learning new tools and developing new applications.

Understanding objects is something that comes easy to some, but more slowly to others. There are several excellent books that can be found in the computer science sections of larger bookstores under object-oriented programming. Some of them are quite readable without a computer science degree. It may help to find someone who is comfortable with the notion of objects and discuss an application. This exercise is often the best way to fully understand the power of object-oriented programming.

ObjectARX is a very powerful tool and is most rewarding to use once mastered. Like most things, it takes time to get used to, and to actually begin to feel comfortable with, but after a while you will wonder how you ever programmed any other way. The same is true with object-oriented programming. Once you really understand it, you will find that you have been more or less doing it all along and now understand why sometimes certain approaches seemed right while others did not.

The following paragraphs give some words of wisdom to programmers who want to learn more and master ObjectARX. Each section is titled by a general background summary. Chances are good that you may fit into one or more so read them all and see if you can choose a path that fits your unique situation.

FOR C++ PROGRAMMERS

If you know the C++ language and have worked in Microsoft Visual C++, then you should have little trouble doing the exercises that are provided in the form of labs in the ObjectARX book. Using these examples, along with the other examples found in the ObjectARX directories, it's not too difficult to find something that is close to what you want. Simply clone that code and use the online help and object browser to locate the exact syntax needed to accomplish the tasks.

Autodesk offers advanced training on ObjectARX. These classes are available through the Autodesk Developer Network. The training classes are useful if you already have some AutoCAD background. If there are enough programmers at your site who need to learn this information, you might also consider hiring a consultant or Autodesk representative to come in and provide more specific training and project advice.

If you do not know how to run AutoCAD, it is strongly recommended that you learn how the system runs and how the interfaces are put together. To accomplish this, take a basic AutoCAD operator training class or work with a tutorial book before getting too deep into ObjectARX. Even though the creation of most applications will not require an intimate understanding of how AutoCAD operates, it does help significantly when communicating with the operators and designers who will be using the software.

FOR AUTOLISP PROGRAMMERS

The C++ programming language contains many standard subroutines that AutoLISP programmers will find familiar. If they are missing, the ADS/ObjectARX library fills in the gaps making it possible to convert AutoLISP programs into C++ programs. The problem is that it's tedious if done by hand. Most of the time it is easier to start from scratch and use the AutoLISP code as a reference since data storage structures can be better utilized. For large AutoLISP applications it is better to convert applications a little at a time, concentrating on those aspects of the system that would benefit the most from the speed enhancements.

There are many reference books on C++ and it is recommended that you find one that is comfortable. Visual C++ does a lot and as a consequence most tutorials will teach you how to do things you may not want to do when planning to write applications for AutoCAD. You may want to concentrate on books that teach the language only (C++), and not those that teach the environment (Visual C++).

The hardest transition for AutoLISP programmers is to get used to the notion of objects containing properties and methods instead of lists of data and functions. Although closely related, and in some programming styles mostly identical, an object can include much more. Through the polymorphism—same name for doing the same thing to different objects—concept extremely succinct code can be the result once the objects are completed.

Visual LISP provides a facility to generate ObjectARX programs but this is not recommended as a tool for creating ObjectARX subroutines for use in AutoLISP. It is strongly recommended as a tool for learning about and working with AutoCAD objects and it provides access to ObjectARX-type functions. From one perspective this would allow for the rapid prototyping of applications in the LISP language. From another perspective this may be all the object level access ever dreamed about in AutoLISP.

FOR VISUAL BASIC PROGRAMMERS

The Visual BASIC language is an elegant tool for working with Windows applications and making them operate together. Using ActiveX automation, formally called OLE, Visual BASIC can call subroutines and functions in different applications and obtain results, or have it carry out operations in the package. AutoCAD supports an ActiveX automation interface that opens the doors to the objects inside AutoCAD. As a result, Visual BASIC is an excellent tool for building applications that serve as a front end to AutoCAD by accepting data in one format and converting it into AutoCAD graphics. The language is also well suited to the creation of reports given an existing drawing in which objects can be identified programmatically or by operator selection.

Unfortunately, both the ActiveX automation and the implementation of VBA found inside AutoCAD do not allow one to define new objects, and are limited to working with existing objects only. In addition, they may not be able to access the objects of a custom application unless the application also supports ActiveX automation. Lastly not all of the events supported in ObjectARX are available to Visual BASIC programmers. What this means is that although Visual BASIC is a great tool for customization, it does not have nearly the power that ObjectARX has in relation to controlling AutoCAD and generating rock solid applications. The question is really whether the application needs the enhanced features found in ObjectARX or can get by with the Visual BASIC tools.

Those who already know Visual BASIC should start with the reference library provided by Autodesk and browse the help files to learn about interfacing with the AutoCAD object system. Using the object browser, and associated help files, one can quickly begin to develop programs that create and read drawings. As the application needs exceed that available in Visual BASIC, it's time to learn how to program in C++ and pick up Visual C++. The development environment is much the same, however, the emphasis on forms is greatly diminished. Much of programming structure already learned in Visual BASIC can be applied in Visual C++.

FOR PROGRAMMING NOVICES

The best place to start programming and working with objects is AutoLISP, as there is no additional investment required except time. Using a tutorial book and a good reference manual, start by reading examples and understanding how they work. Then progress into writing small routines. Tutorial or learning books on AutoLISP will provide good exercises for you to do in that regard. There are two versions of AutoLISP from which to chose currently. The "traditional" AutoLISP is not as powerful and therefore not as complex as Visual AutoLISP, however, an effort should be made to use the latter. Visual AutoLISP also has the ability to create ObjectARX programs so that if you decide to migrate towards ObjectARX programming in the future, there is a great learning tool already waiting.

If you plan on learning C++ too, see the AutoLISP programmers' suggestions above for moving from AutoLISP into the C++ language. It is not a difficult step, although there are major differences in the way programs are presented to the computer. The main advantage of learning AutoLISP is that it can be done without the purchase of other tools, with exception of a book or two, and you can experiment easily right at the command line in the AutoCAD system.

AutoLISP provides all the needed tools to manipulate a drawing and create expert systems of all sorts of designs. But like the Visual BASIC options, it cannot create a new object and it has no access to AutoCAD events whatsoever. The next step from AutoLISP would be towards Visual LISP or straight into C++ with ObjectARX. Many applications can be solved with just the LISP language provided they do not also need to interface with objects outside of AutoCAD. One should keep in mind that no matter how one started, the ultimate goal would be to learn C++ and utilize the power of ObjectARX to get the customization done best.

OBJECTARX RESOURCES

There are several excellent resources for learning more about the ObjectARX programming environment. These resources require that one is familiar with the Visual C++ system and knows how to activate the help system and the object browser. If you do not have the ObjectARX programming information from Autodesk, you can get it by joining the Autodesk Developer Network. The developer network provides regular updates of AutoCAD software, including pre-release versions for developers to get a jump-start on the next version. Although this program costs several hundred dollars a year to join, it is a good investment for serious AutoCAD developers. More information about the developer network is available directly from Autodesk or on their web site at www.autodesk.com.

When you get the ObjectARX development kit it includes a CD full of programs and examples. These examples are very useful when learning the system and finding out how to employ certain features of the programming language or library. There is a series of examples found in a "lab" or "classlab" directory which is meant to be used in a training class. Students are expected to fill in the blanks with the missing code and make the problem work. These make for excellent prototypes from which to build new applications exploiting the same concepts.

The online help system provided with ObjectARX requires that you know the names of the functions to get the best possible information. This is why tables were provided in this book listing several of the names found in the libraries. Another way to learn these names is to use the object browser in the Visual C++ workbench. The object browser will show all the names of the objects defined in the various classes. Select one and press the F1 key to see more details about it. Although this may seem cumbersome to describe, it does allow one to float around the library and become comfortable with the naming scheme used.

Lastly, you might find a local consultant or developer that can serve as a sounding board for ObjectARX programming ideas. Before contracting a programmer it is a good idea to have a firm understanding of the problem and how it can be solved. If you don't already have this, and are able to define it in writing, then you may need to contract a system analyst who understands the tools found in AutoCAD, as well as an understanding of the application at hand.

The important thing to keep in mind when learning ObjectARX is that although it can at first appear overwhelming, it really is not. There is a large library of tools that can be used, and more than one programmer has been confused as to which is the best to use for any given circumstance. The best way to learn is to try it and see what happens. User feedback and looking at other programs provide excellent mechanisms to improve as they demonstrate what is needed or can be done with the tools at hand. Looking at other software that do new things will inspire a good programmer to think of new ways to deliver an application. Above all, my advice is to keep on programming.

SUMMARY

The differences between command-driven and object-driven applications was discussed at the start of this chapter. That led into a more detailed study of the programming languages available for customizing AutoCAD using objects as we looked at ObjectARX, Visual BASIC, and AutoLISP from an object-oriented point of view.

This chapter then explored some conceptual examples of how objects can be employed in the world of CAD/CAM/CAE applications development. Although no details were presented, some ideas were tossed around relating the concepts of engineering problem solving to the tools made available through object-oriented programming. Finally some guidance was provided as to where to learn more depending on the background of the reader. If you have read this far, your journey is just beginning!

CHAPTER 9

Building an ObjectARX Application

Writing programs for ObjectARX is only part of creating an application. After creating source code files the next step is to convert them into a compiled, executable program. Microsoft Visual C++™ Versions 5.0 and 6.0 is used for the creation of ObjectARX modules for AutoCAD (Release 14 and AutoCAD 2000). You must have the same version of the compiler and the ObjectARX CD installed on your hard drive in order to follow the steps in this chapter.

THE RIGHT TOOL FOR THE JOB

Various companies provide numerous compilers to select from when looking for C++ language solutions. Although other environments work, the C++ compiler recommended in AutoCAD Release 14 ObjectARX documentation is Microsoft Visual C++ Version 4.2b. It has been thoroughly tested by Autodesk for use as an ObjectARX development platform and is recommended over any other version. If you have Microsoft Visual C++ Version 4.2, download the 4.2b patch at the Microsoft web site. We will use Microsoft 5.0 for the ObjectARX application in AutoCAD Release 14. The same steps as used in 5.0 are also used in 6.0 to create ObjectARX modules for Release 14. It is not necessary to have 5.0 and 6.0 installed on your programming machine if you are developing for both platforms.

To develop applications for AutoCAD 2000 using ObjectARX you must have a copy of the ObjectARX 2000 library files as well as Microsoft Visual C++ Version 6.0, or later. Previous versions of Microsoft Visual C++ have not been tested. Instead we used the tool kit suggested by Autodesk that includes Microsoft Visual C++ 6.0 with Service Pack 2 installed.

The Microsoft Visual C++ Version 6.0 development environment improves on an already good platform by providing more features. There are deeper technical issues that make this version of C++ the one to use in conjunction with AutoCAD and ObjectARX 2000. Version 5.0 does not do the job in a consistent manner. If you do not have a subscription update, then get the shelf update from your Microsoft dealer.

Future ObjectARX and AutoCAD releases may have different compiler requirements. In past AutoCAD releases, C/C++ programmers were required to update their programming tools to match them. Because Autodesk does not supply the C/C++ tools there are lags between version support for products that are migrating at their own speeds. It is not unusual for advanced systems such as AutoCAD to require older versions of compilers with which to build integrated environments. Substantial amounts of development have taken place while the older version was the most current. When this complicated system was released, the programming tools experienced a new release in the same time frame.

The price paid for the performance and capability of ObjectARX is that you must invest in a compiler tool, more if you intend to work under multiple platforms. Programmers find the features well worth the price. The cost of the C++ compiler and linking system is about $500. It is also recommended that the Windows® NT operating system be used because of its superior error trapping during development. Windows 95 will work fine under most circumstances but will lock-up when major problems occur. Windows® 98 has not been tested for this purpose.

THE FILES INVOLVED

An ObjectARX program contains more than C++ source code files. There are header, definition, and resource files, plus libraries. The first thing to understand when developing programs in this environment is what all these files contain and exactly how they must be set up. (See Table 9.1)

When writing a program, start with a CPP file. It contains application source codes and can be manipulated using any text editor. The Microsoft Visual C++ developer environment contains a great text editor for this purpose. Make use of the online help and object browser options to speed the programming process along. C++ keywords and functions are color coded—as are the comments—making for easy program code reading.

For ObjectARX applications developers it is strongly recommended that a standard startup CPP source file be created in which you simply fill in the blanks—or augment—to build the starting portions of the application. A minimal standard source file contains the entry point definition routine and associated initialization functions, along with the basic header files called for in include statements. A minimal start-up CPP file is presented in the section titled STARTER.CPP.

When running Microsoft Visual C++ there will be other files created: object files, libraries, workspaces, and files used by the compiler and linker. Most are of little or no importance to the programmer. The files you will work with the most are the CPP (C++ source) and H (header) files. This is why separate directories are recommended for each project for proper file management.

Extension	Purpose
CPP	C++ Source file containing program codes, calls to ARX routines, and whatever else is needed for the application. A project can have more than one CPP file associated with it.
H	Header Source file containing definitions of variables, classes, and constants. A project can have multiple header files to define the elements used in the project.
DEF	Module Definitions file containing attributes that describe the DLL (ARX) module.
RC	Resource file used when working with the MFC tools in Visual C++.
LIB	Library file that contains binary modules to be linked for building the application. Supplied by Autodesk.
DSP	Project file, created by Microsoft Visual C++.
DSW	Project workspace, created by Microsoft Visual C++.
ARX	Runtime extension file, the output that results when an ObjectARX program has been successfully built.

Table 9.1 *Project File Types*

STARTER.CPP

The following contains a minimal start up CPP (C++) source file to be used to develop more tools. If you are learning C++, remember that characters that start with ac, ads, arx, and rx are from the ObjectARX library. Comments start with double slash marks (//).

Header inclusions are at the start of the source file. Header files define data types and function parameter sequences used by the programs. Several header files are defined in each module to cover the types of data used as well as the ObjectARX interface details.

The below list shows header files included in my starter C++ file. Those that are not needed are either ignored or commented out by placing the double slashes in front of the #include statement.

```
// Standard header files
#include <adesk.h> //set up data types
#include <rxdefs.h> //define ACRX structure
#include <aced.h> //edit services, new commands
```

```
//#include <adslib.h> //Release 14 ADS components
#include <adsdlg.h> //dialog box control information
#include <dbents.h> //entity definitions
#include <dbsymtb.h> //symbol table definitions
#include <dbmain.h> //database manipulations
#include <rxobject.h> //custom object
#include <acgi.h> //drawing functions
#include <rxregsvc.h> //registration service
//#include <adsdef.h> //traditional ADS interfaces
```

An application does not need all the header files shown above. Only a few may be required. For the minimum, an AutoCAD Release 14 based application should include the aced.h and adslib.h headers. ObjectARX 2000 applications need the adesk.h, aced.h, and rxdefs.h header files. There are additional header files required by other functions such as the boundary and region handling routines. Consult online help for the easiest way to learn which particular header is required.

You can also elect to leave the headers in place since all they do is define function prototypes and macros. They will not greatly increase the size of the output file, which will range from four kilobytes on up.

If the application makes use of global variables and functions, a header file is the most convenient mechanism for maintaining them for access by the other functions involved.

INITAPP()—INITIALIZE THE APPLICATION

With the AutoCAD data type definitions handled in the header files the next step in the STARTER.CPP file is to define a function to be called when ObjectARX is started. This is not required, but is strongly recommended as a matter of programming style. It is called by the entry point function when the initialization command is given. The application defines new objects, sets up system level control opportunities, and registers new commands. In order to promote readability and future source code expansion, setup statements are confined to a single subroutine. It is defined before the others in the CPP source file and only called once in the main entry point function.

When this function is run, AutoCAD may or may not be fully functional or have a drawing loaded in the editor. Do not run AutoCAD commands or expect AutoCAD to respond to anything related to the command or event processors. This entry defines internal objects and establishes basic reactor links and functions that need only to be executed once when the application loads. When an ObjectARX program is loaded AutoCAD calls it twice. The first time is when the ObjectARX program is loaded from the disk and ready to run. The second follows after the drawing is ready, which will be immediately if the drawing is already loaded. Any time a new drawing is loaded the second service request code will be sent to the entry point. The initialization service request is only made when the program is initially loading into the system.

```
// Initialization function
void initApp()
{
// Register commands and classes
//
acedRegCmds->addCommand("MY_COMMANDS",
    //Command group
"GLOBAL", //Global name
"LOCAL", //Local name
ACRX_CMD_MODAL, //Flags
FuncName //Function to run
);
}
```

The above example registers a dummy command. Change the parameters to match the application or copy it for additional commands. The command group allows you to reference all commands at a future time as a singular object.

Global and local names of the commands permit applications to support localized versions. The local name can be in a language such as German or French while the global name can be the command in English. How the local and global command names are used will vary depending on the native language of the developer and target users.

Several flags are available for commands being added to the AutoCAD command stack. Most of the time, the command will be defined as modal, meaning it runs by itself. Options also exist for the creation of transparent commands.

The last parameter to the addCommand function is the name of the function to run when the command is activated. This function is from your own code. The value supplied as the parameter is not a string, it is instead the symbol name of the function. The symbol name will cause C++ to send the address of the symbol. For this to work, function name and parameters must be defined in the source code before being referenced in the function call to add a new command. The function being called can be stored in a different file that is part of the project. This does not remove the requirement that the function be declared before referenced elsewhere. This requirement is from the C++ language.

UNLOADAPP()—UNLOAD THE OBJECTARX ROUTINE

If your program adds new objects or commands, or changes AutoCAD, it should undo these operations when the exit request is delivered. Like the initialization routine, this function is not required but highly recommended for modularity and readability.

The routine below removes commands from the command stack when the application is unloaded. Failure to do this means an operator can still run commands while the program is no longer loaded. When commands are run without a supporting program the result can be a disaster. Since the command stack simply stores pointers, what will happen will be based on what is stored at that memory location.

```
// Remove commands and classes
void unloadApp()
{
// Remove the command group(s)
acedRegCmds->removeGroup("MY_COMMANDS");
}
```

By having all new commands added as a single group, you remove them all at once from the command stack with the macro call shown above. Most developers name the command group after the application. In the generic STARTER.CPP file the name "MY_COMMANDS" was used.

ACRXENTRYPOINT()—THE MAIN ROUTINE

The main routine is where the program execution starts. With an ObjectARX program it is the function acrxEntryPoint(). Two parameters are supplied to the entry point function. One is a message code deciphered by the program and then acted upon. The other is a pointer to a common memory area used for data transfer.

In most applications, the main routine consists of a single switch statement that tests the value of the message code. The message code is an integer that is one of the options in the RXDEFS.H header file. Instead of using integer values—which are subject to change, even though the probability of that event is unlikely—use the names supplied by ObjectARX. They improve code readability. Message names are listed in Chapter 4.

An important item to keep in mind is that an ObjectARX program will receive two messages as it loads:

The first indicates the application is being initialized. At this stage the program declares new objects and defines new commands. AutoCAD may not be fully functional so a drawing may not be loaded at the time the first request is made. As a result, user and drawing interaction is strongly discouraged. There are housekeeping tasks that occur when the first message is received by the entry point module. The application needs to unlock itself so it can be unloaded. This is optional, however, most applications will want to allow the unload function to operate. The second task is to declare your program as being multiple document aware. This step is only taken in ObjectARX 2000—and later—as multiple document interfacing was not supported in AutoCAD Release 14 and earlier. Multiple document processing issues are discussed later.

The second message is given after a drawing is loaded. During this time ObjectARX programs define AutoLISP links using acutDefun() or ads_defun() for AutoCAD Release 14) and establish database linkages. Reactors to existing entities can be set up during this message call to the entry point.

If another drawing is loaded, the second request will reoccur. However, the first request is only made while the application is being loaded into memory. When defining new commands and objects they do not need to be redefined when a new drawing loads. What needs redefining are any transient relationships. They are items such as reactors to database objects, ownership chains, and resetting values of custom objects to defaults.

```
// Entry point or "main" function
extern "C" AcRx::AppRetCode
acrxEntryPoint(AcRx::AppMsgCode msg, void* pkt)
{
 switch (msg) {
 case AcRx::kInitAppMsg: //First request
   acrxDynamicLinker->unlockApplication(pkt);
   acrxDynamicLinker->registerAppMDIAware(pkt);
   initApp();
   break;
 case AcRx::kUnloadAppMsg: //Exit request
   unloadApp();
   break;
 case AcRx::kLoadDwgMsg: //2nd load request
   break;
 case AcRx::kUnloadDwgMsg: //Unloading drawing
   break;
 case AcRx::kSaveMsg: //drawing is being saved
   break;
```

```
    default:
    //other options in RXDEFS.H
    break;
    }
    return AcRx::kRetOK;
}
```

Programming for other messages will depend on the application. Save information to your own data files at the same time AutoCAD does or load the data from a data file when a drawing loads. Most applications do not do a great deal when these events occur so it is strongly advised you keep application activities to a minimum for speed purposes. Operators are sensitive to issues such as the time required to save a drawing or load a new one.

If you migrate from ADS to ObjectARX you will most likely make use of another message value not shown in this sample routine. The acutDefun() function results in the kInvkSubrMsg message being sent. The application calls ads_funcode() to retrieve the function code and then processes accordingly. The function integer code is the same as in ADS. It is assigned to the function name for AutoLISP to use when the acutDefun() function is employed. The switch statement used to process the function code in an original ADS program moves to a subroutine and gets called when the invoke subroutine message is received. Migrating from ADS to ObjectARX is simple if you want to continue with the AutoLISP based interface.

Any other entries into your custom application are based on commands registered and reactors established within the program itself. The main entry point exists only to service the standard messages that AutoCAD will send to the ObjectARX program.

The main entry point is the systems administrator for your ObjectARX program. It sets up initial objects, handles the definition of application entries and exits, sets up the main details with AutoCAD, and establishes memory allocations as needed by the application. Other functions, which are linked through the command stack and object reactors, are the other entry points into the program. They handle the real application.

THE DEF FILE

ObjectARX uses the contents of the Definitions File (DEF) to link with AutoCAD. The DEF file contains names used by the C++ linker when laying out a dynamic link library (DLL). DEF contains the name of the application as seen by other applications. It is important to link the proper names so that AutoCAD interfaces with the ObjectARX program.

DEF files are included in the project when you define the DLL for compilation. AutoCAD uses this mechanism to converse with the ObjectARX programs and may change in future ObjectARX releases.

Group	What it Contains
LIBRARY	The name of the DLL file with an extension.
EXPORTS	List of names for the function to export. These are the functions that other tasks can link into. AutoCAD requires three names to be present. They are acrxEntryPoint, _SetacrxPtp, and acrxGetApiVersion. AutoCAD uses these three functions when loading the DLL to initiate conversation.
DESCRIPTION	An optional description of the library and its purpose.

Table 9.2 *DEF File Contents*

A typical DEF file appears as follows:

> **LIBRARY myStuff**
>
> **EXPORTS acrxEntryPoint**
>
> **_SetacrxPtp**
>
> **acrxGetApiVersion**
>
> **DESCRIPTION 'My ObjectARX program is very fast'**

Usually, the only change in the DEF file is the name of the library and the description.

Keep a standard DEF file handy for use when starting new projects. When you start the project, copy the standard DEF file into the project directory and change the name to match. Edit the values in the file and attach it to the project.

ONE AUTOCAD, ONE OBJECTARX

Although it may be convenient to think that the ObjectARX application is loaded and attached to each individual drawing, this is not the case. Even though previous versions of ObjectARX could open multiple drawings, only one was considered the current drawing and thus the main drawing being manipulated. Actually, ObjectARX applications are loaded once and then attached to the AutoCAD program itself. The ObjectARX module has its own variables and, like AutoCAD, is responsible for keeping track of which drawing it is addressing at any given time. It also means that new commands resulting from the ObjectARX program load will be available in all open drawings as long as the ObjectARX module remains loaded.

If your application requires different variables for each drawing that is loaded then you will require a buffer area for each. There are several ways to accomplish the buffering of data for a drawing and all are triggered by the same event. When the drawing is loaded, the application receives a message through the entry point module indicated such as (kLoadDwgMsg). When the message is sent, your ObjectARX module defines a new instance of a class that houses all variables for that drawing such as short cuts to the tables and entities of interest. When input is requested from the user, the drawing status is checked and appropriate action taken with the proper variable settings. This aspect of MDE coding is what will stop many ObjectARX modules from being ported from Release 14 to AutoCAD 2000, as there will need to be some rethinking of the procedures.

What is important to remember is once the ObjectARX module is loaded, all drawings can access it. If there are new commands defined, they will be available to all the drawings. If there are AutoLISP functions defined in the ObjectARX module—through the acedDefun subroutine—then they are available in the other open documents as well. With an import request, these same external function definitions are available to Visual LISP modules too. If the ObjectARX module is unloaded while a drawing is active, it will no longer be available to other open drawings.

The ObjectARX application is notified of new drawings loading through the kLoadDwgMsg message sent to the main entry point. It is relatively easy for your program to keep track of what is in memory. The only difficulty is when asking the operator to make a selection or choose a point. Your application must be aware of what drawing it is talking with.

ARE YOU AWARE?

To make an application aware of multiple documents you tell AutoCAD your system is MDI (multiple document interface) aware. From the AutoCAD program's perspective, that is all that is needed and AutoCAD will consider the program able to handle multiple document switches that take place as a natural consequence of running AutoCAD 2000. That brings us to the second aspect of making an ObjectARX application MDI aware—is it really aware?

To be truly aware of the multiple document mode you introduce more checks on the input supplied. If your program asks a Yes or No question via the acedGetKWord—this used to be ads_getKWord in Release 14—function, a prompt will appear on the command line and the operator can freely switch between different drawings in the view area. But does it matter in that case? For a simple question, the answer would be not really. If the application were asking if a save was required, make sure you have a pointer to the right drawing in your own variables. If the save question could be related to any drawing selected, then make sure you get the current drawing linkage before performing the save method in the drawing object.

For your program to be aware of changes in the current document selection, more testing and coding is needed. As another illustration a program might request a point input from the user and that point may be considered relative to another point. If a different drawing is selected, the input would be meaningless in the context of the question being asked. What should an application do under these circumstances? The choices are basic, but may require substantial coding to achieve. The first choice would be to force the user back to the drawing you want them to select from and request another input. The next option would be to ask the user if they want to abort the application. A third way would be to accept the input. There may be a new way to use your software not considered previously.

LETTING AUTOCAD KNOW YOU ARE AWARE

To inform AutoCAD 2000 that your ObjectARX module is MDI aware there is a utility function that must be called when the module is first loaded. Use it when your application is first loaded and the entry point module receives the kInitAppMsg message. By telling AutoCAD 2000 that your program is MDI aware you are telling the system to proceed in MDI mode if already there.

If your program does not tell AutoCAD it is MDI aware by calling the required utility function, then AutoCAD will assume that the ObjectARX module is not MDI aware and wants to run in Single Document Mode (SDI) instead. An error message displays on the command line and the system reverts to SDI mode. The error message and sudden change in operation is uncomfortable to users if it comes as a surprise.

To make your program MDI aware add the C++ function call to the source code.

AcrxDynamicLinker->registerAppMDIAware(pkt);

The parameter pkt is the data packet found in the entry-point subroutine parameter list. It is used to keep track of the AutoCAD system internally and should not be modified by your application. Only use it when needed.

Insert the code module into the entry-point subroutine in the middle of the switch statement that is processing messages sent to the application. When kInitAppMsg is received as the ObjectARX program is loaded and starts to run, these sort of housekeeping tasks need to take place. They involve functions needed to establish a link with AutoCAD. They include defining new command groups, unlocking the application so it can be unloaded, and informing AutoCAD your program is ready for the multiple document environment.

OBJECTARX DEMONSTRATION PROJECT FOR
AUTOCAD RELEASE 14

This section is a tutorial where readers create a new program file. You must have Microsoft Visual C++ 5.0, AutoCAD Release 14 (14.01), and ObjectARX Version 2.x libraries.

If you do not have these tools available—or manage programmers who use them—this section will introduce exactly what is technically required when writing an ObjectARX program using Visual C++.

Bulleted items are the required steps to accomplish the essential tasks.

The demonstration project also shows how to set up the Visual C++ programming environment to develop ObjectARX programs. ObjectARX programs are dynamic link libraries (DLL). They are treated differently because AutoCAD controls their activities. As a result of these differences, you apply minor changes to the default setup parameters for both the C++ compiler and system linker.

- In Windows NT or Windows 95 start the Microsoft Visual C++ 5.0 developer studio.

ObjectARX programming takes place inside the Developer Studio provided with the Microsoft Visual C++ 5.0 package.

This is a new project. The first step is to define where to house the project. Project control files keep track of various source and compiled object files. They also define the basic parameters for compilation and linking.

- In the developer studio select the [File] -> [New] pull-down menu options.

A dialog box presents new file options to create. There are multiple file types to create using this utility.

- Click the Project tab.

A list of possible project types will appear (See Figure 9.1). ObjectARX programs are 32-bit style Dynamic Link Libraries. As we create the new project we also define where we would like to store the project. Fill in the entire dialog box.

- On the Projects list select Win32 Dynamic Link Library.

- In the directory type where you will be building the project. I use a directory named \PROJECTS.

- Type the project name. The directory will change as you type the project name.

- Click OK to create the new project file setup.

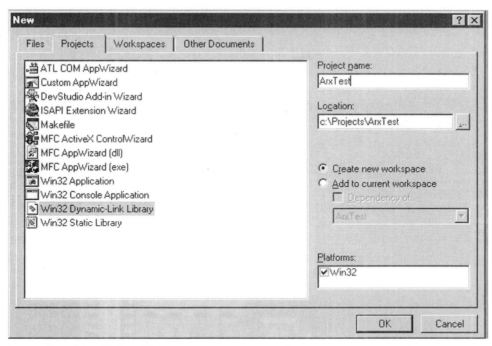

Figure 9.1 *New Project Options*

If the directory already exists, it will be used. Otherwise a new directory matching the name specified will be created. Several files will be created in the project directory. They are related to the control of the source and project status. The only file of interest is the DSP file. It contains project workspace information. It is of interest because it is the file selected when you want to reload a project for later edit work. For the most part, the new files can be ignored. The Developer Studio and Visual C++ are using them.

If you have created the source files add them to the project. In most cases, now is the time to write program source codes into text files. The files will be CPP (C++ source) and H (header source) files. We are not going to create a header file for this demonstration, but we will create a C++ source file.

- In the developer studio select the [File] -> [New] menu options.

- In the dialog box select the Files tab to view options for creating new files.

- In the list of available file types select "C++ Source File."

- In the "File name" edit box type the name of the new file. You do not need to supply an extension; it will automatically be CPP. Figure 9.2 shows the screen dialog box at the end of this operation.

- To create the new source file click OK.

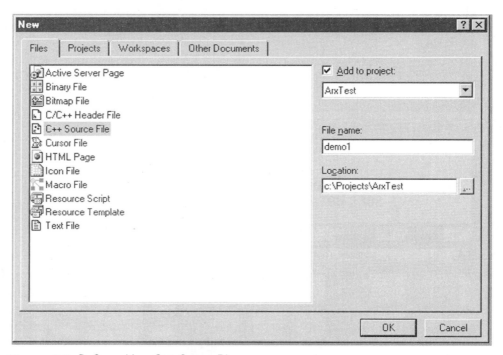

Figure 9.2 *Define a New C++ Source File*

Click OK and the file is automatically added to the current project and stored in the project directory.

A Visual C++ program is verbose when compared to other languages such as AutoLISP or Visual BASIC. There is a much greater degree of control over the environment in which your program will run when using Visual C++. There are functions for accessing remote corners of the operating system. When the ObjectARX library is added, that same intimate level of access is also granted. The result is the curse of long variable names—to avoid conflicts with the many other names available—and strange character combinations. Like other languages, Visual C++ and ObjectARX take time to get used to.

The following listing is a minimal C++ program to demonstrate the compiler set up. If you do not have an ObjectARX C++ program to test and would like to test the compiler and ObjectARX system, the following provides enough code to get started. If you do not like to type source code, use the code provided with the ObjectARX Version 2.0 for the sample labs.

A single command is defined called "HOWDY." It runs the function ObjectARX_GREETINGS that prints a message in the text window. When the ObjectARX application is unloaded, the command group is removed from the command stack. This program demonstrates the basic principles in building an ObjectARX program.

The initialization and unload functions discussed previously have been removed for your convenience. If you intend to expand this minimal module then those code enhancements—the initApp and unloadApp—should be incorporated into the program code accordingly.

At the start of the program code header files are included. The selected headers are those needed when programming most ObjectARX applications. Following the include statements is the function that runs when the command is typed at the command line. Next in the code is the main ObjectARX program entry point. It receives two parameters but only the first is used in coding. The message code—provided in the variable msg—is tested to see what action is requested of our program. Use the names from the AutoCAD headers—AcRx::kInitAppMsg—and switch expression tests for various possibilities. This program checks for two possible message requests, a program load and an unload. When program load is detected, a new command is added to the AutoCAD command stack. This new command is linked to the function defined above by name. When an unload request is received, the command group is removed from the AutoCAD command stack.

- Type the following list into the CPP source file. Match every character exactly. Upper and lower case is important in the Visual C++ language.

```
#include <adesk.h>
#include <rxdefs.h>
#include <aced.h>
#include <adslib.h>
void ObjectARX_Greetings ()
{
 ads_printf("\nHello and Howdy from ObjectARX!");
}
//
```

```
extern "C" AcRx::AppRetCode
acrxEntryPoint(AcRx::AppMsgCode msg, void* pkt)
{
switch (msg) {
case AcRx::kInitAppMsg:
  acrxDynamicLinker->unlockApplication(pkt);
  acedRegCmds->addCommand("MY_COMMANDS",
  "HI",
  "HOWDY",
  ACRX_CMD_MODAL,
  ObjectARX_Greetings
  );
  break;
case AcRx::kUnloadAppMsg: //Exit request
    ads_printf("\nCommand group removal");
  acedRegCmds->removeGroup("MY_COMMANDS");
  break;
default:
  //other options in RXDEFS.H
  break;
  }
return AcRx::kRetOK;
}
```

Type in the function and verify syntax, then build the definitions file. Create a new Text file named ArxTest.DEF. Use a template file stored on the system that can be loaded into the text editor and saved as a new name after changing the library and description. This demonstration application is labeled ArxTest. Name the definitions file accordingly.

- Create a new file or open a template and name it ArxTest.DEF.

- Type the following into the DEF file using the text editor.

```
LIBRARY ArxTest
EXPORTS acrxEntryPoint
 _SetacrxPtp
```

acrxGetApiVersion

DESCRIPTION 'A first ObjectARX program test'

The program-coding portion is complete. Make sure the text matches exactly. Upper and lower case are very important in the Visual C++ environment.

Now, build the project. To create a program we compile the C++ code into binary object files and then merge them with library files. The output program is a dynamic link library suitable to run inside AutoCAD. In order to make it suitable for AutoCAD there are changes that must be made in the project to match Release 14 ObjectARX interfaces.

- In developer studio select Project then Settings.

The Project Settings dialog box appears (See Figure 9.3). It has numerous tabs on the right side used to navigate to the various aspects of setting up a Visual C++ project build. Change values in the C/C++ and Link tab areas.

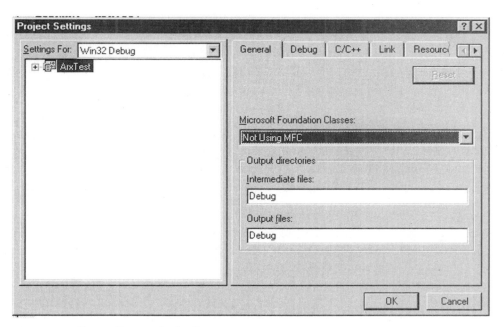

Figure 9.3 *Project Settings Dialog Box*

Within each tab are several categories of information panels. When moving from one area of this dialog box to the next, the most common approach is to select the tab, then select a category within its grouping. In some cases more than one category changes.

- Next to Settings, click the pop-up list then select All Configurations so that the debug and release versions have the updates applied.

Unless you plan to use the Visual C++ debug mode select the Win32 Release option. Use the debugger with ObjectARX programs to step through the break points you establish in your code. More specific information on how to run the debugger in conjunction with ObjectARX programs is found in the online help files provided with ObjectARX modules.

- Select the C++ tab to reveal the C++ compiler settings.

- On the Category pop-up list select the Code Generation to show the settings available under that particular topic.

- Change Use Runtime Library to Multithreaded DLL. The display now matches Figure 9.3.

That is the only change required in the code generation panel. Next, the preprocessor needs to be informed of some minor additions.

- On the Category pop up list select Preprocessor.

- Under the entry for Preprocessor definitions add ACRXAPP, RADPACK.

- If you do not plan to do a lot of ObjectARX programming then the ObjectARX include file directory needs to be added to the list of additional directories to search. By default this will be \ObjectARX\Inc.

Figure 9.4 shows the dialog box after changes have been made. The \ObjectARX\Inc directory was not part of the dialog box because we will add the reference to the include directory to the system defaults.

C/C++ portions of the ObjectARX program project are ready. The last step is to update linker information so the libraries are properly integrated into the application.

A linker program takes object files created by a compiler and merges them with library files of additional binary objects to create an executable program. ObjectARX requires some subtle changes to the standard dynamic link library to accommodate operations inside Release 14. Different releases of the ObjectARX interface require different settings. This effect can be seen in the next chapter in which AutoCAD 2000 requirements are discussed.

- In the Project Settings dialog box select Link to reveal link options.

- Type in the program name and the extension ObjectARX. If no extension is provided, the system will automatically append DLL. This can be changed later with no consequence to program operation. The display now matches Figure 9.5.

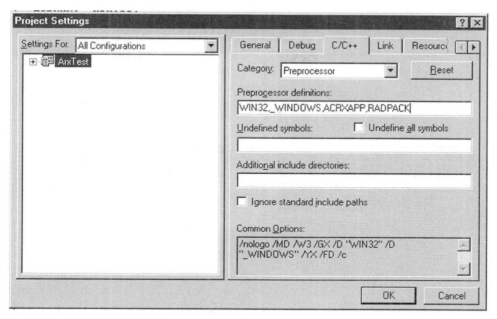

Figure 9.4 *Preprocessor Definition Changes*

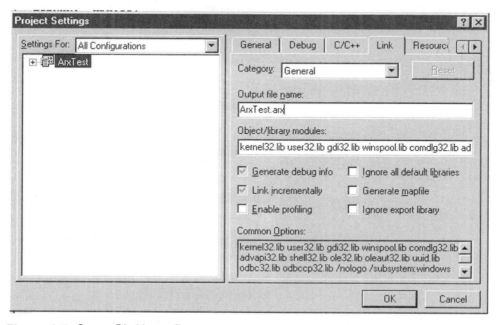

Figure 9.5 *Output File Name Change*

- Click the Category pop-up list, then select Output in order to work with the computer storage and location characters of the output module.

In the Base address field type 0x1c000000. With the exception of the 'x' and 'c' characters, these are numeric values. The number is a hexadecimal value.

In the Entry-point symbol field type DllEntryPoint@12. All characters are alphabetic except "12" at the end of the string. (See Figure 9.6)

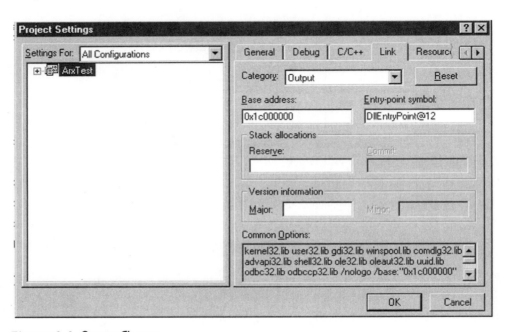

Figure 9.6 *Output Changes*

When the dialog box matches Figure 9.6, click OK to accept all changes.

The next steps establish system defaults. Once done, you will not have to do them again since the system defaults are saved for future programming sessions.

- Add the ObjectARX include directory to the search set. Click Tools pull-down menu then select Options. A dialog box will display with multiple tabs. Click the Directories tab.

- Under the Show Directories for prompt select Include files.

- Select the button for a new entry into the list. It is shown as a dotted file folder icon.

- Type in the directory name \ObjectARX\INC or select the triple dots to use the file explorer to locate the directory.

- Apply the changes by clicking OK.

The system defaults to the ObjectARX include directories. You will not have to repeat the above steps for future program modules.

ObjectARX libraries are located in the \ObjectARX\Lib directory. These files were installed when the ObjectARX CD was installed.

The easiest way to add the libraries is attach them to the project. Most projects require the RXAPI.LIB and ACAD.LIB files. If you add new commands you need to map in the ACEDAPI.LIB file. The following steps show how libraries are added on.

- From the Projects pull-down menu click Add to project, then select Files. A dialog box will display for different files to be added.

- Select Library files (LIB) as the type of files to be displayed.

- Navigate to the \ObjectARX\LIB directory.

- Select the required library files. It is recommended you start with RXAPI.LIB, ACEDAPI.LIB, and ACAD.LIB. Add others as needed. Figure 9.7 shows the library file select taking place. Use the Control-Pick combination to select multiple libraries.

- Click OK to have the files added to the project.

The ObjectARX program and environment are ready for building. These changes and manipulations are done once per project and, in most cases, you will only add new files as the project expands.

- On the Build pull-down menu select Build ArxTest.arx. Press F7 to begin the building process.

You now have an ObjectARX file in either the debug or release directory. To navigate between them use the Build pull down menu the select Set Active Configuration. A dialog box is displayed showing options between debug and release. Unless you have set break points in your program, use the release configuration when generating the output file.

To run ObjectARX, start AutoCAD Release 14 and click APPLOAD. In APPLOAD, select the Files and locate your ObjectARX module. Then select Load to load the ObjectARX program into memory. To test it, on the command line type HOWDY. For the same result you can also type HI.

Although it may seem complicated to build an ObjectARX program there are only a couple changes from the base default method of building a dynamic link library.

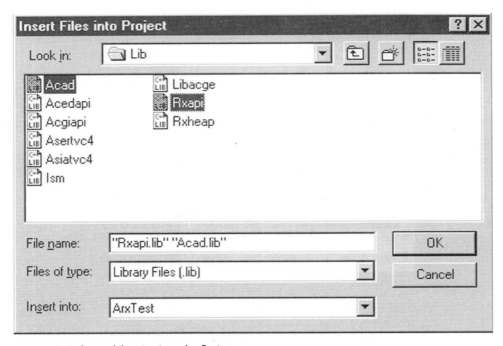

Figure 9.7 *Insert Libraries into the Project*

Several steps are involved in setting up an ObjectARX program but are done once for any given project. Once prepared continue to compile/link and test over and over again without changing settings.

The changes applied to a default 32-bit dynamic link library to build an ObjectARX file are as follows.

1. Make use of multithreaded DLL.

2. Add ACRXAPP, RADPACK to the preprocessor definitions.

3. Set the base address to 0x1c000000

4. Set the entry point to DllEntryPoint@12

Otherwise, the creation of an ObjectARX program is identical to that of a normal DLL. To make use of MFC and other programming interfaces, add their required changes to your programs as normal.

OBJECTARX DEMONSTRATION PROJECT FOR AUTOCAD 2000

This section serves as a tutorial. To create a new program file you must have Microsoft Visual C++ (Version 6.0), AutoCAD 2000, and ObjectARX 2000.

If you do not have these tools available—or manage programmers who use them—this will introduce you to what is technically required when writing an ObjectARX 2000 program using Visual C++ 6.

Bulleted items are the required steps to accomplish the essential tasks.

The demonstration project also shows how to set up the Visual C++ programming environment to develop ObjectARX programs. Although ObjectARX programs are DLL, they are treated differently because AutoCAD controls them. As a result, you must apply minor changes to default setup parameters for the C++ compiler and system linker.

  • **Start the Microsoft Visual C++ 6.0 developer studio in Windows NT, Windows 95 or Windows 98.**

ObjectARX 2000 programming takes place inside the Developer Studio provided with Microsoft Visual C++ 6.0.

First define where to house the project. Project control files keep track of various source and compiled object files and define the basic parameters for compilation and linking.

  • **In the developer studio select the [File] -> [New] pull-down menu options.**

A dialog box presents the new file options to be created. Multiple file types can be created using this utility.

  • **Select the Project tab.**

A list of possible project types appears. ObjectARX programs are 32-bit style Dynamic Link Libraries. Create the new project and define where to store the project on the disk system.

  • **Select Windows 32 Dynamic Link Library.**

  • **Type the project name (in this example arxtest2)**

  • **Type the project location (in this example C:\Projects\arxtest2)**

The New dialog box appears as seen in Figure 9.8.

  • **Click OK to move to the application Wizard associated with the creation of a Win32 Dynamic Link Library.**

  • **Click Finish, then in the next dialog box click OK for the Wizard to complete project initialization.**

We are building an empty DLL project. We will supply all code and resources as needed to build it. Try other options and more files are created than needed for the majority of ObjectARX development tasks.

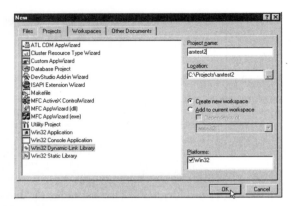

Figure 9.8 *New Project Options*

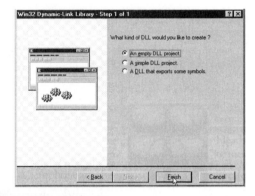

Figure 9.9 *Wizard Completion Dialog Box*

If you make use of the MFC dialog components—and not the DCL system—then you must build a more complex project as resources are added. To make use of the Microsoft Foundation Classes use the MFC application wizard to create the initial project, then change the MAIN entry to the acrxEntryPoint name.

For this example an empty DLL project will suffice. After the wizard completes initial DLL definitions there will be several files created that are parts of the project that we will not edit or change. The Developer Studio uses them when building the application project.

The next step is to define C++ source code. The Visual C++ Developer Studio is used to define your source code.

- In the developer studio, select the [File]-> [New] menu options.

- From the New dialog box select the Files tab to view options for creating new files.

- From the list of available file types select C++ Source File.

- In the File name edit box type the name of the new file. You do not need to supply the extension; it will automatically be CPP.

- Click OK to create the new source file.

Click OK and the new file is added to the project directory and file tree. It is empty, waiting for your input. One aspect of C++ programming that frightens some people is that the code is verbose when compared to other languages such as AutoLISP or Visual BASIC. The reason is that you have more control over the computer environment in C++ than you do in other languages. Visual C++ provides functions for accessing the most remote corners in the Windows operating system and ObjectARX has the same intimate relationship with AutoCAD.

The result of such power is the curse of long variable names—to avoid conflicts with the other available names—and sometimes strange looking combinations of characters. Like other languages, Visual C++ and ObjectARX take some time to get used to.

Now add source code for your application. The following list needs to be added into the code windows of the newly created CPP file. When typing in this source code, every character must be exact. In C++, upper and lower case are different, thus the name AcDb is not the same as ACDB.

- Type the following into the code window for the file arxtest2.cpp.

```cpp
#include <adesk.h>
#include <rxdefs.h>
#include <aced.h>
void ObjectARX_Greetings ()
{
 acutPrintf("\nHello and Howdy from ObjectARX
    2000!");
}
extern "C" AcRx::AppRetCode
acrxEntryPoint(AcRx::AppMsgCode msg, void* pkt)
{
 switch (msg) {
 case AcRx::kInitAppMsg:
     acrxDynamicLinker->unlockApplication(pkt);
```

```
            acrxDynamicLinker->
        registerAppMDIAware(pkt);
            acedRegCmds->addCommand(
        "MY_COMMANDS",
            "HI",
        "HOWDY",

        ACRX_CMD_MODAL,

        ObjectARX_Greetings
        );
        break;
    case AcRx::kUnloadAppMsg:
        acutPrintf("\nCommand group removal");
        acedRegCmds->
        removeGroup("MY_COMMANDS");
        break;
        default:
         break;
        } //end switch
        return AcRx::kRetOK;

}
```

- After typing in the file contents save the file.

Next a definitions file needs to be defined.

- Create a new text file named arxtest2.def. Here is where a template file helps as in the STARTER.DEF file explained earlier.

- Type the following into the arxtest2.def file.

```
LIBRARY ArxTest2
EXPORTS acrxEntryPoint
 _SetacrxPtp
 acrxGetApiVersion
DESCRIPTION 'Greetings to the programmer'
```

Save the definition file. Make sure the upper and lower case usage of the text matches.

This completes the coding phase. More advanced applications will have more code to be entered and this part of the development will be longer.

The next steps need only be applied once per project. They will result in changes to the way the source code is treated when you request the project be built. Building a project involves compiling the C++ code into machine object code then linking that with the libraries to be used.

- In the Developer Studio select the [Project] then [Settings] menu options.

A multiple tabbed dialog box will appear with all project settings. The left side of the dialog box contains configuration selections for them. Select All Configurations for this example as well as the majority of standard ObjectARX applications.

Visual C++ provides a debug and release configuration option. The debug configuration is used with your own applications development and its usage is outside of the scope of this primer book. See Microsoft C++ documentation for assistance in using the debugger in conjunction with C++. The basic concept is that AutoCAD is the host application for the DLL you are creating and the system will need to know how to launch the program. AutoCAD will start and stop as a result and the debugger link will not appear.

Back to our simple example. We change the way the C++ defaults were set up by the empty DLL project. Because our application will be running inside AutoCAD, we need to make some minor adjustments to project settings.

- Select the C++ tab to reveal project control options for the translation of C++ code into object format.
- In Category select Code Generation.
- Change the Use Runtime Library selection to Multithreaded DLL.

That completes the only change to the code generation. Next we add a preprocessor directive that enables our program to link with AutoCAD.

- In the Category pop up menu in the C++ tab select Preprocessor.
- Under the entry for Preprocessor Definitions add ACRXAPP.
- If you do not plan to do a lot of ObjectARX programming the ObjectARX includes file directory needs to be added to the list of additional directories to search. By default this is \ObjectARX\Inc.

The C++ portions of our ObjectARX program are ready. The last step is to update linker information so the libraries are properly integrated into the application.

A linker program takes the object files created by a compiler and merges them with library files of additional binary objects to create an executable program. ObjectARX

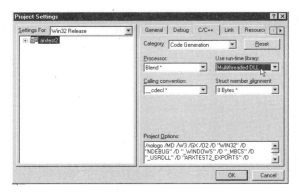

Figure 9.10 *Code Generation Changes*

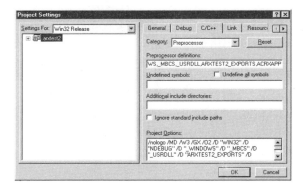

Figure 9.11 *Preprocessor Changes*

requires some subtle changes to the standard dynamic link library to accommodate operations inside AutoCAD 2000.

- Select the Link tab to reveal link options.

- Type the name of the program and the extension ObjectARX. If no extension is provided, the linker will automatically append DLL. This can be changed at a later time with no consequence to the program generated.

- On the Link tab in the Category pop-up list select Output. This selection will cause the computer storage location data for your program to appear.

- In the Base Address field type 0x1c000000. With the exception of the 'x' and 'c' characters, these are numeric values. The number is a hexadecimal value.

We have changed the base address where the entry point for the DLL will reside. This is an AutoCAD requirement. If you do not do this, the ObjectARX program will not link with AutoCAD when we attempt to load it.

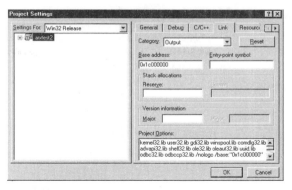

Figure 9.12 *Link Output Changes*

- Click OK to accept all changes made thus far.

The next steps establish system defaults. Once done, you do not have to do them again in other projects since the system defaults will be saved for future programming sessions.

- Add the ObjectARX include directory to the search set. On the Tools pull-down menu select Options. A dialog box will display with multiple tabs. Select the Directories tab.

- On the Show Directories for prompt, select Include files.

- Select the button for a new entry into the list. It is shown as a dotted file folder icon.

- Type in the directory name \ObjectARX\INC or select the triple dots to use the file explorer to locate the directory.

- Click OK.

The system now defaults to the ObjectARX include directories. You will not have to repeat the above steps for future program modules.

ObjectARX libraries are located in the \ObjectARX\Lib directory. These library files were installed when the ObjectARX CD was installed.

The easiest way to add the libraries is attach them to the project. Most projects only require the acad.lib, acedapi.lib, acrx15.lib, acutil15.lib and rxapi.lib files be included. The following steps show how libraries are added on.

- From the Projects pull-down menu click Add to project, then select Files. A dialog box will display for different files to be added to the project.

- Select Library files (LIB) as the type of files to be displayed.

- Navigate to the \ObjectARX\LIB directory.

- Select the library files required. It is recommended you start with RXAPI.LIB, ACEDAPI.LIB, ACUTIL15.LIB, ACRX15.LIB and ACAD.LIB. Add others as needed by the application. Use the Control-Pick combination to select multiple libraries.

- Click OK.

The ObjectARX project is now ready to build your first ObjectARX module under ObjectARX 2000. These changes may seem to be excessive but they are done once per project and, in most cases, you will only add new files as the project expands.

- On the Build pull-down menu select Build ArxTest.arx. Or press F7 to begin the building process.

Provided everything is correct, you now have an ObjectARX file sitting in either the project, debug, or release directory. To navigate between the two (Note: aren't there three instead of two?), use the Build pull-down menu Set Active Configuration option. A dialog box is displayed showing the options between debug and release. Unless you have set break points in your program, use the release configuration when generating the output file.

To run the ObjectARX program, start AutoCAD 2000 and use the ObjectARX command to load the ObjectARX file. After loading, on the , command line type Hi. The program will respond with the text string Hello and Howdy from ObjectARX 2000.

Although at first glance it looks complicated to build an ObjectARX program, there are only a few changes from the based default method of building a dynamic link library. There are several steps involved in setting up an ObjectARX program but they are only done once for any given project. Once prepared you continue to compile/link and test over and over without changing any settings.

The changes applied to a default 32-bit dynamic link library to build an ObjectARX 2000 file are:

1. Multithreaded DLL in code generation of C++ tab.

2. Add ACXAPP to preprocessor definitions.

3. Set the base address to 0x1c000000

4. Change the output name to include the ObjectARX extension.

SUMMARY

This chapter presented a basic walk through the tools and basics of building a first ObjectARX application. If you were successful in building the application then you are well on your way to building very advanced applications that will tightly integrate with AutoCAD. The steps presented in this chapter are the ones I've used over and over again to successfully write ObjectARX application programs. There are many other approaches that can be taken to achieve working programs and the versions presented in this chapter represent one of the simpler styles.

Good luck in your programming efforts, whether you are writing code or managing a large project. Learning more about AutoCAD and ObjectARX should be easier with the foundations presented in this book. ObjectARX is a deep subject that has many interesting twists and turns that can yield tremendous power to your applications. An understanding of how ObjectARX works produces great insights into how other applications and AutoCAD itself can be improved.